Using and Applying Mathematics at Key Stage 1

Using and Applying Mathematics at Key Stage 1

A Guide to Teaching Problem Solving and Thinking Skills

Elaine Sellars and Sue Lowndes

Routledge
Taylor & Francis Group
LONDON AND NEW YORK

First published 2003 by David Fulton Publishers Ltd

Published 2016 by Routledge
2 Park Square, Milton Park, Abingdon, Oxfordshire OX14 4RN
605 Third Avenue, New York, NY 10017

First issued in hardback 2016

Routledge is an imprint of the Taylor & Francis Group, an informa business

British Library Cataloguing in Publication Data
A catalogue record for this book is available from the British Library.

ISBN 13: 978-1-138-17577-8 (hbk)
ISBN 13: 978-1-85346-961-9 (pbk)

Typeset in Great Britain by Keyset Composition, Colchester, Essex

Contents

Introduction

What are the characteristics of a mathematically able child?

- Abstract thinking
- Logical reasoning
- Application – being able to transfer skills from one area to another
- Independent thinking – being able to see things from a different perspective
- Ability to work independently
- Organisational skills
- Understanding new concepts with ease and being able to use them
- Relatively rapid mastery of mathematical techniques, knowledge and habits
- Good spatial perception
- Mathematical curiosity
- Mathematical memory of a higher order
- Ability to verbalise their thoughts; flexibility of thought; originality of thought
- Ability to find mathematical shortcuts
- Perseverance. Broad attention spans which enable them to concentrate and persevere in problem solving
- Analytical thinking – ability to evaluate, analyse and generalise

This list is not exhaustive but gives a good feel for what we know as teachers makes a child able, before we ever test them to try to prove and quantify that ability.

As teachers, we know that to teach effectively we need to start from where the child is, and then build on their current skills and knowledge. A key problem, however, is that many current teaching practices, which able pupils meet every day in their classrooms, ignore the above list. The nature and structure of the current curriculum means that there is very little time for considering how we might develop the innate abilities that mathematically able children possess. Much of what we teach is about content rather than the thinking process.

Look at the list again. Where does it say that an able child is born with an innate ability to 'add together $\frac{1}{2}$ and $\frac{1}{3}$'? It doesn't of course; the list is not content based. Yet at the moment we focus almost exclusively on teaching mathematical methods and knowledge rather than on helping children to improve the innate abilities that make them able. How will they reach their potential without good teaching?

Maths is about how our brains work and think. When do we teach all our children, regardless of their ability, strategies to improve how they think and problem-solve?

Brilliant mathematicians throughout history have had mentors. Where are their mentors in school? Maths is a sociable subject; it is something that mathematicians do together. At a maths conference you see mathematicians doing maths, not just talking about it.

Unfortunately many of the mathematical tasks given to able pupils actually work against the abilities in the list above. For example, how will able children ever build up their stamina and perseverance if they are only ever given tasks which take seconds or at best a few minutes to complete? This is what this book is about.

Our philosophy

- The majority of able children can and should be taught with their peers in order to provide for their social and emotional needs as well as mathematical needs. This requires tasks that work in mixed ability classrooms. Only a tiny percentage of pupils require a different approach.
- Able children have a right to direct teaching in the same way that all other children do. The teacher role is pivotal in providing the necessary learning environment.
- In the classroom, being able at maths does not equal being quick at maths, i.e. if an able child is working on a task that is appropriate to them they will not necessarily take less time to complete their work than other pupils. In fact it may take them longer. Able children will not always be the first to finish.
- Practical equipment is not just for those who find maths difficult. Plenty of spatial puzzles must be provided for all pupils.
- Able pupils should be given work appropriate to their level and not be made to plough through work that is unnecessary. Also their reward for a piece of work done well should not be another worksheet; more of the same to keep them busy and quiet.
- Able pupils need to be taught strategies to enable their thinking skills to progress. Teachers would not dream of starting a fraction topic off by adding $\frac{1}{2}+\frac{1}{3}$ – there would be a great deal of preliminary work before this. Yet, it is quite common to find that there is no progression built in to the teaching of the using and applying part of the curriculum. Teachers and pupils often dive straight in; there is no coherence or progression from task to task and never from year group to year group.
- A sustained piece of work that requires perseverance, logical strategies, refinement of methods and extension of the original task is not the same as a 10/20-minute back-of-the-envelope type problem. Both types of problem solving need to be taught.
- The key to this is teaching. Problem solving has to be taught even though pupils are exploring maths at the same time.
- Relying purely on acceleration to meet the needs of an able child is not the best way to meet those needs. This often stores up problems for the future. There is more maths out there than that which is contained in the numeracy strategy. The numeracy strategy is a good base but an able child requires breadth and depth in their knowledge and understanding. Able pupils often require decelerating not accelerating. They need to be stretched sideways, not pushed on. Just because they can do a piece of content from Year 6 when they are in Year 4 does not mean that it is necessarily appropriate to

teach it then. Maths is a huge jigsaw puzzle. Without some of the pieces, a child may be able to guess what the picture is but they will not see the real beauty of maths. If we only think about accelerating pupils then we deny them that beauty.

The two books in this series, for Key Stage 1 and for Key Stage 2, can each be used with whole classes. They provide a basis for a structured curriculum for the teaching of problem solving. Eventually teachers will supplement these tasks to develop a broad and enriched curriculum.

Each tried and tested investigation is clearly explained, with ideas on how to introduce the task to a class. Full solutions, resource sheets and examples of children's work, where appropriate, are included. Any of these pages can be photocopied for use within your own class or school.

We recommend that you try to work through four to five of the tasks with your class each year. Instead of using the 'spare week' in the timetable to play catch up, try doing this instead.

Successful teaching relies on well planned lessons. You do need to work through the tasks before giving them to your class to work on. Remember that mathematics is a sociable subject; it works best when done in collaboration with others. You could try to get thirty minutes of the weekly staff meeting set aside to work on a task together. It will be much more fun than what you usually do!

We would be very pleased to hear from you if you have any comments or questions on the activities in the books, or if you'd like to share your experiences of using the activities in school. You can contact us by email on esltrain@btinternet.com

1 ROCKET TO THE MOON

The task

Play this as a game:

Two players need two 1–6 dice between them and one rocket sheet.

Each takes turns to roll the dice. The two numbers obtained are added together.

They then colour that number in on the rocket (or cover with a counter),

The winner is the one with the most numbers coloured in.

> !! The rocket deliberately contains the numbers zero and one. This is to allow the children to discover for themselves that they cannot make these numbers.

Introducing the task to your class

Once the children have played the game a number of times ask them some questions.

- Could they cover 0 or 1? If not, why not?
- Did some numbers come up more often than others did? Were some numbers hard to get?

Build on this by changing the rules. When the dice are rolled this time, place counters on the numbers attained. When the game is finished ask the pupils to record how many counters are on each number. Collect their results altogether.

- What do they notice?
- Why is this happening?

Writing up the investigation

Ask each pupil to record all the possible throws. You will want to encourage systematic listing. Some of the pupils may need the dice in front of then to record the possible throws. This can be done in a variety of ways; below are two possible ways.

Die 1

	+	1	2	3	4	5	6
	1	2	3	4	5	6	7
	2	3	4	5	6	7	8
Die 2	**3**	4	5	6	7	8	9
	4	5	6	7	8	9	10
	5	6	7	8	9	10	11
	6	7	8	9	10	11	12

possible throws						score	no. of ways
1,1						2	1
1,2	2,1					3	2
1,3	2,2	3,1				4	3
1,4	2,3	3,2	4,1			5	4
1,5	2,4	3,3	4,2	5,1		6	5
1,6	2,5	3,4	4,3	5,2	6,1	7	6
	2,6	3,5	4,4	5,3	6,2	8	5
		3,6	4,5	5,4	6,3	9	4
			4,6	5,5	6,4	10	3
				5,6	6,5	11	2
					6,6	12	1

As the teacher you could choose to set up the recording system in this way or let the children experiment. They will find it hard to know if they have missed out any combinations unless they have a system.

Extension ideas

- Use 0–5 dice
- Use different shaped dice
- Use a combination of 0–5 and 1–6 dice

Ask the pupils to predict what is going to happen before they play the game.

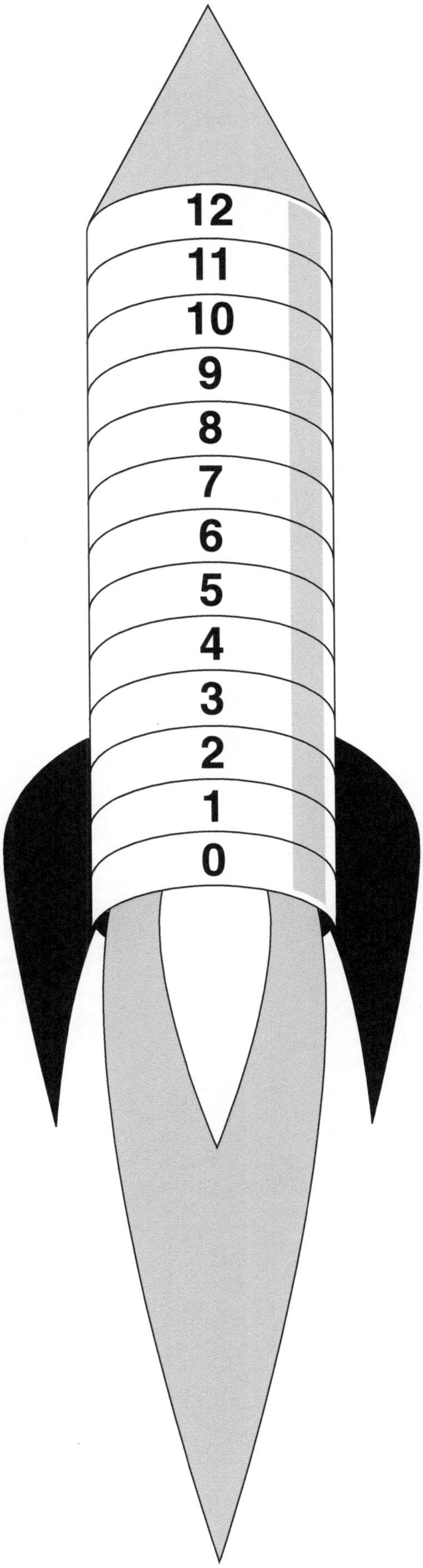
12
11
10
9
8
7
6
5
4
3
2
1
0

2 MAKING 10p

The task

A task to encourage systematic listing.

Introducing the task to your class

You need lots of 1p, 2p, 5p, and 10p coins.

Ask the pupils to list the ways to make 10p, either on paper or using coins.

Pupils may record by placing the coins under plain paper and rubbing over them with a crayon, or using money stamps, or by listing.

Ask them all to start with 10p.

The two possible routes are now to change this for 10 × 1p coins or 2 × 5p coins.

If they choose the 1p route then their next choice should progress from this and not leapfrog to, for example, 2 × 5p.

A better choice from 10 × 1p coins would be to swap 2 × 1p coins for 1 × 2p coin and build on this. Gradually increase the coin denominations. Then consider the 5p combinations, again being systematic.

> !! Of course, for some of your pupils you will be pleased that they can do any combinations, but here we are considering able pupils and how to teach them the basic building blocks of problem solving of which systematic working is one.

Writing up the investigation

Systematic listing for making 10p:

1. 

2. 

3. 

4 

5. 

Changing list no. 3:

Changing list no. 4:

Changing list no. 5:

3 TABLES AND CHAIRS

The task

A task to help pupils to work systematically and to spot patterns. Encourage pupils to discuss their ideas.

Introducing the task to your class

Start the activity with the whole class. Use either real tables and chairs in the classroom or square tiles and small blocks on their classroom tables. Ensure that **all** children use the equipment to simulate the arrangements before they attempt to write anything down. Always refer back to the practical equipment to overcome difficulties.

Writing up the investigation

Included here are two possible types of worksheets. They vary depending on age and ability. You may wish to adapt these to suit your own class (pp. 16 and 17–18).

Pupils will need to have a clear idea of what you expect them to write down if you wish to use the less structured sheet. For this reason if 'tables and chairs' is one of the first investigations you do, it will help if all pupils work through the pack where they write on the sheets.

This will also provide their parents with a clear picture of the steps involved in problem solving. This should mean that parents could support this learning more effectively at home.

!! Children with poor reading may need targeted support.

Teaching points

Reinforce the systematic way the problem is solved.

First we gather information, then we summarise this information in a table of results. We look for patterns and make predictions. **Praise mistakes**. Children need to feel confident in experimenting, and getting things wrong is vital to improve pupils' skills in refining and improving their problem-solving strategies.

Pupils' work **must** be a chronological report of what they have done. For instance, they should not draw a table of results until they have some results to put into it! Using the printed sheets can show pupils the various steps they need to follow. Consider using pens rather than pencils for writing and do not allow rubbing out. Pupils should not write up their work in best. At first their work will be messy but with practice this will soon improve.

For your able pupils, encourage them to write down any number patterns they see. Initially do not demand full sentences; try to get their thoughts written down when they think them.

If this is too difficult you can ask them to use a symbol to indicate that they have an idea:

Or a thought bubble

and then when you see the symbol you can ask them what they were thinking and perhaps write it down for them yourself.

Extension ideas

Ask pupils to relate the pattern to the physical arrangement of chairs. For example, in the arrangement

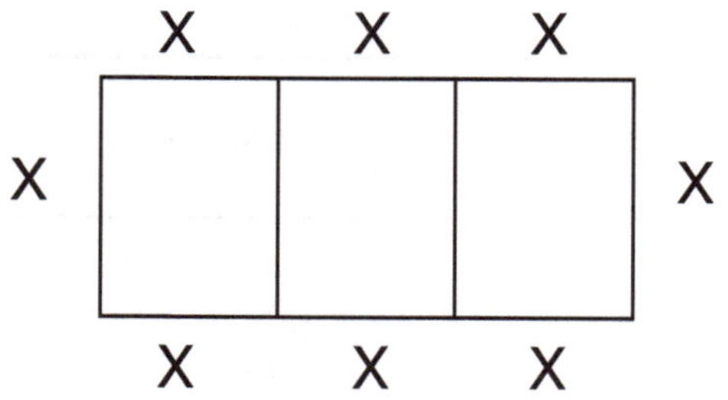

they may see that you double the number of tables then add 2. Can they explain why this works from the diagrams?

Build on this for your able pupils by introducing this arrangement:

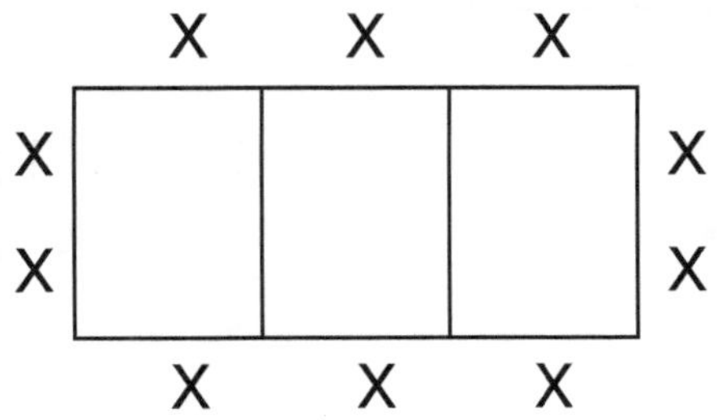

Ask them to predict how to find the number of chairs required by using what they have learnt from the previous example.

What about this arrangement . . . and so on:

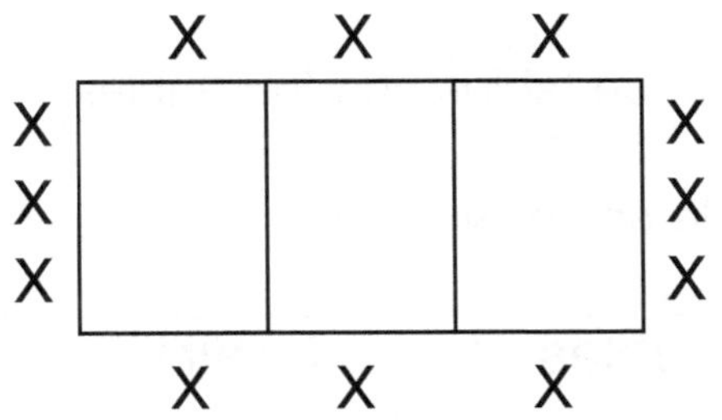

Can your able children write a sentence formula?

e.g. 'the number of tables is doubled to get the number of chairs'
'the number of tables times 2 gives the number of chairs'

You may need to scribe for them initially and/or dictate a suitable rule if you realise that they have got it but can't quite verbalise it.

You may wish to write a few alternatives on the board and see if they can pick out the one that works.

Again you are introducing them to mathematical language. Some of your pupils may feel happy with '$t \times 2 = c$' if you explain what the letters stand for.

!! remember that '*t*' represents the **number** of tables not just tables.

Remember you are still teaching them even though they are problem solving.

Tables and Chairs

A school dining room uses square tables.

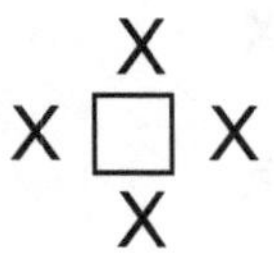

Key: □ table
X chair

Four chairs will fit around one table. Copy this diagram.

Draw two tables and record how many chairs will fit around the two tables, like these:

Now draw the diagrams for three tables and four tables.

Record your results in a table as shown below.

Tables t	Chairs c
1	4
2	
3	
4	

Can you see any patterns in your results? If not, try a few more examples.
Can you find a rule?

How many chairs would fit around 6 tables, 10 tables, and 20 tables?

Further investigations

How many chairs would fit around the tables if they were arranged as shown below?

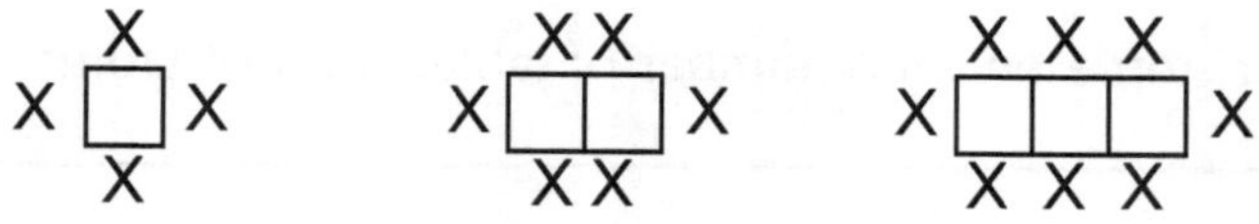

Now investigate rectangular tables.

Name: ______________________________

Tables and Chairs

The Task

It is dinner time. We need to put the chairs at the tables.

We have to do it properly.

Like this:

X □ X   1 table has ____________ chairs

X □ X

2 tables have ____________ chairs

X □ X

X □ X

X □ X   3 tables have ____________ chairs

X □ X

Draw the chairs.

Use X for a chair.

4 tables have ____________ chairs

Draw 5 tables and the chairs.

5 tables have ____________ chairs

We can show our answers like this:

tables	chairs
1	2
2	4
3	
4	
5	
6	
7	
8	

Can you write in the missing numbers of chairs?

You may want to draw pictures.

Or you may know the answer.

Can you see a pattern in the numbers?

Write it down.

____

____

____

Do you know how many chairs you need?

(a) 3 tables need _______ chairs

(b) 7 tables need _______ chairs

(c) 10 tables need _______ chairs

(d) 50 tables need _______ chairs

If you have 6 chairs how many tables are there? _________

X X X X X X

If you have 12 chairs how many tables are there? _________

More tables and chairs

More people come in for dinner.
We need to do the tables and chairs in a different way.

Like this:

1 table

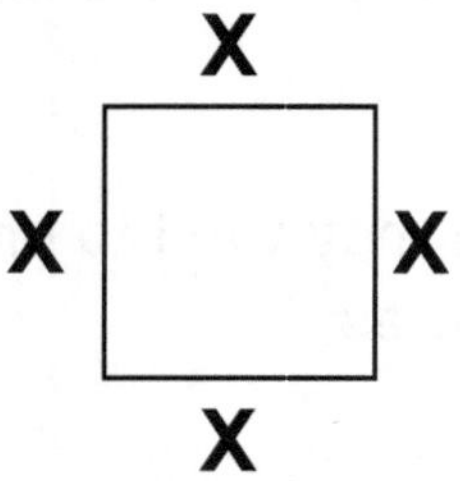

2 tables

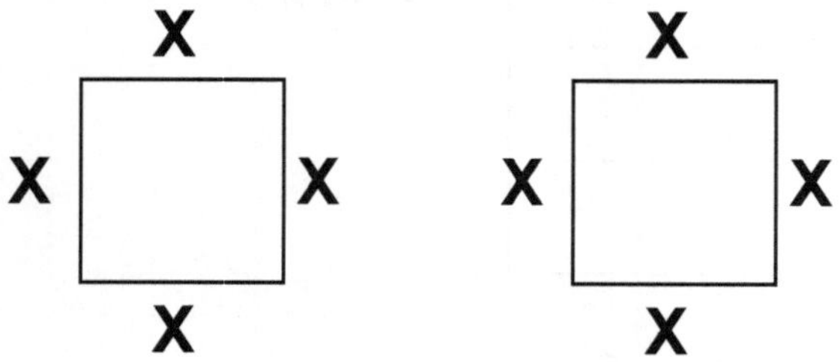

3 tables

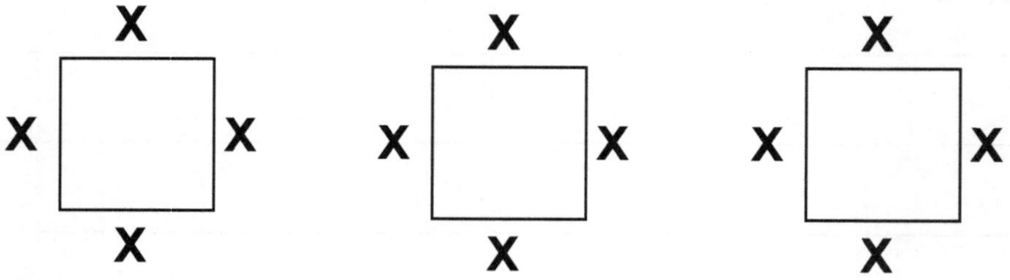

Fill in the numbers of chairs.

tables	chairs
1	4
2	
3	
4	
5	

You can draw pictures or you can work out the answer.

Can you see a pattern in the numbers? ____________________

Even more tables and chairs

We can push the tables together.

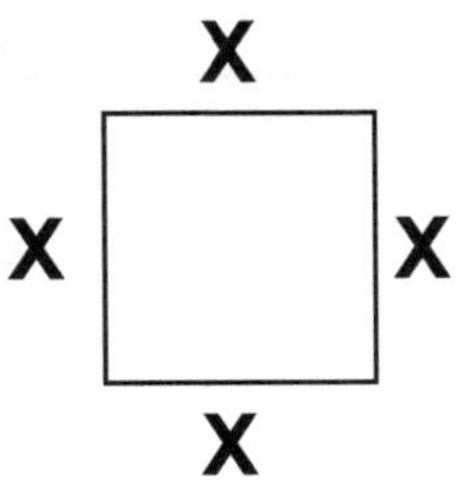

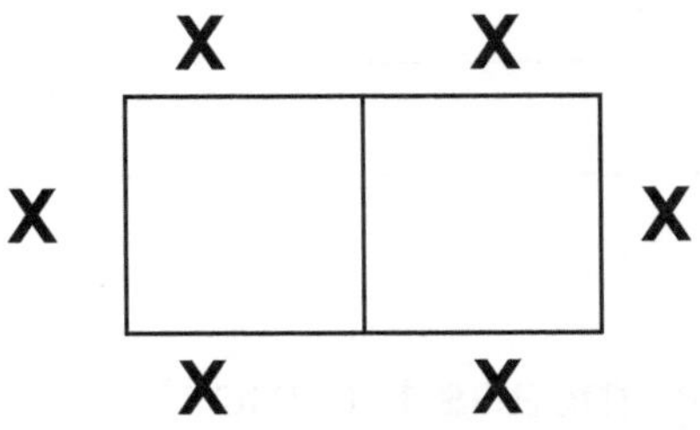

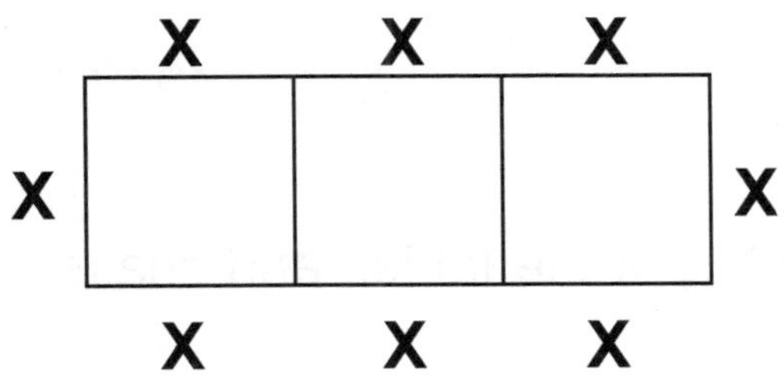

Put the **X** in these pictures:

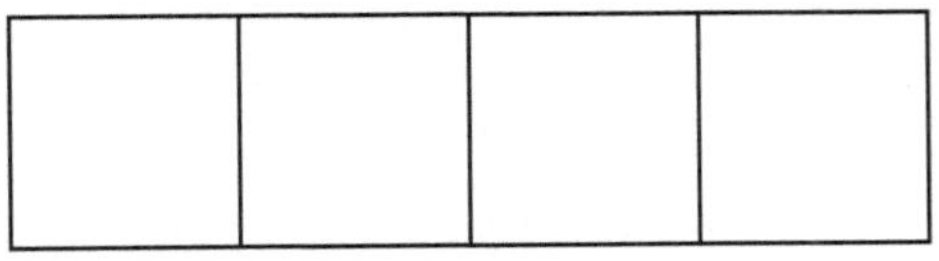

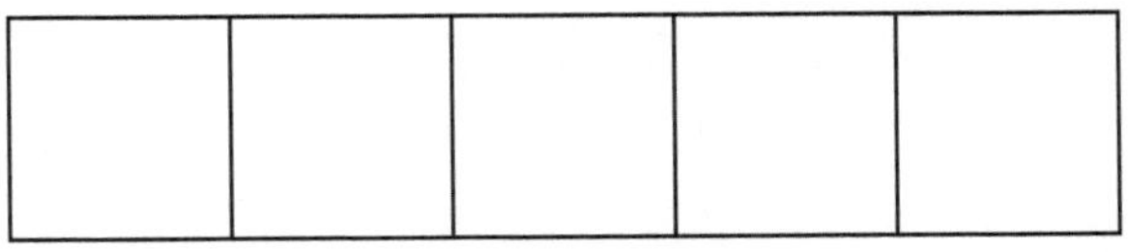

Fill in the numbers of chairs.

tables	chairs
1	4
2	
3	
4	
5	

We have used three different ways to put the chairs at the tables.

Which way do you like best? Why do you like it?

Which way means the most people can sit down?

Can you think of a different way to put the tables and chairs?

Draw it.

Year 1 Name: ALex

Tables and Chairs

The task

It is dinner time. We need to put the chairs at the tables.

We have to do it properly.

Like this:

1 table has 2 chairs

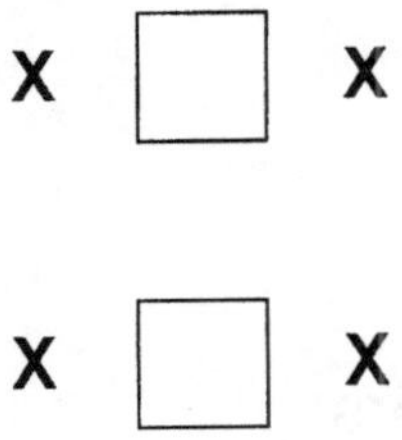

2 tables have 4 chairs

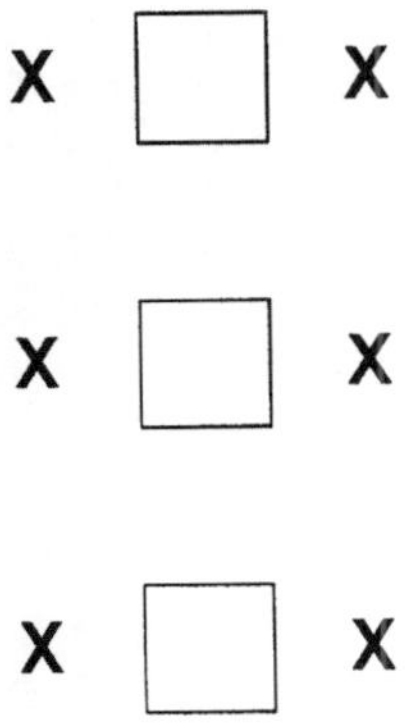

3 tables have 6 chairs

Draw the chairs.

Use X for a chair.

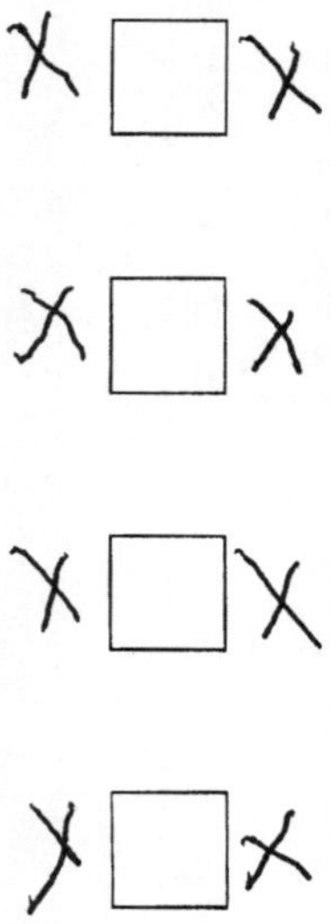

4 tables have 8 chairs

Draw 5 tables and the chairs.

5 tables have 10 chairs

We can show our answers like this:

tables	chairs
1	2
2	4
3	6
4	8
5	10
6	12
7	14
8	16

Can you write in the missing numbers of chairs?

You may want to draw pictures.

Or you may know the answer.

Can you see a pattern in the numbers?

Write it down. Its In cows and dubls

Do you know how many chairs you need?

(a) 3 tables need 6 chairs

(b) 7 tables need 14 chairs

(c) 10 tables need 20 chairs

(d) 50 tables need 100 chairs

If you have 6 chairs how many tables are there? 3

x x x x x x

If you have 12 chairs how many tables are there? 6

More tables and chairs

More people come in for dinner.
We need to do the tables and chairs in a different way.

Like this:

1 table

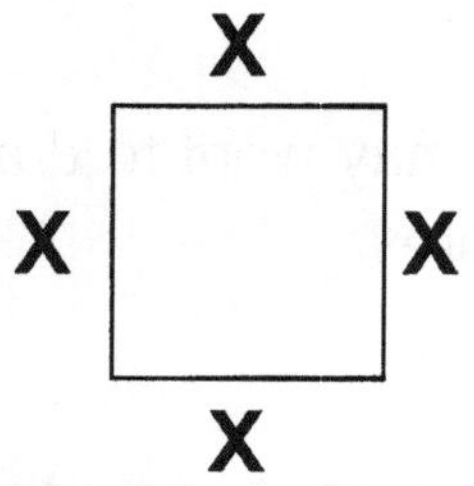

2 tables

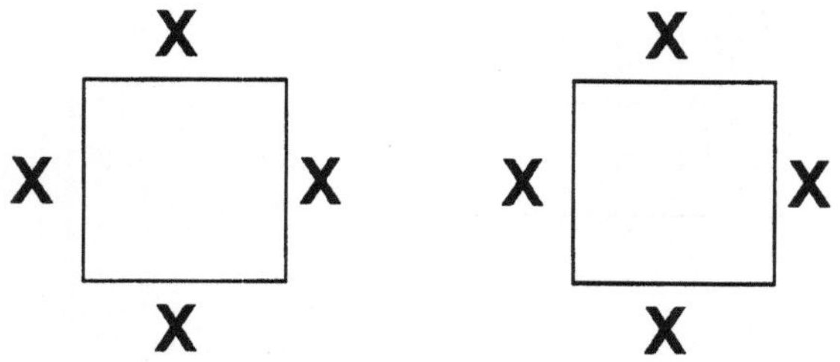

3 tables

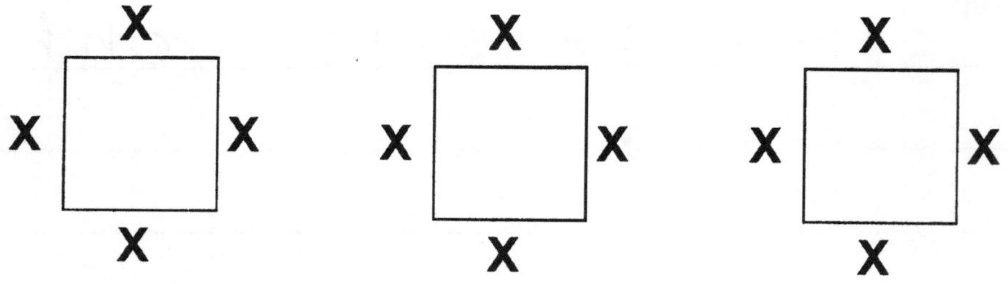

Fill in the numbers of chairs.

tables	chairs
1	4
2	~~4~~ 8
3	12
4	16
5	20

You can draw pictures or you can work out the answer.

Can you see a pattern in the numbers? It is in 4s

Even more tables and chairs

We can push the tables together.

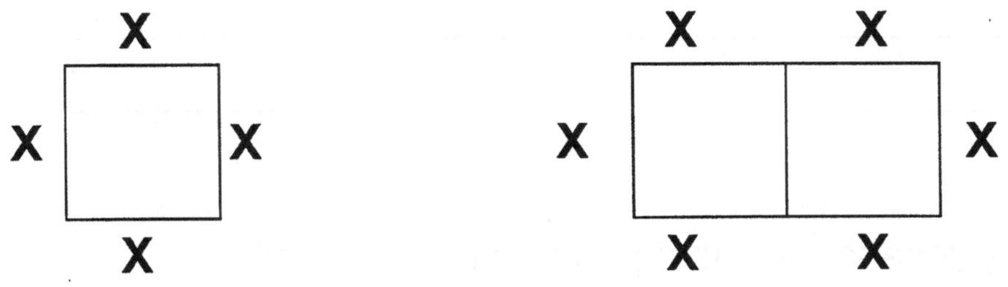

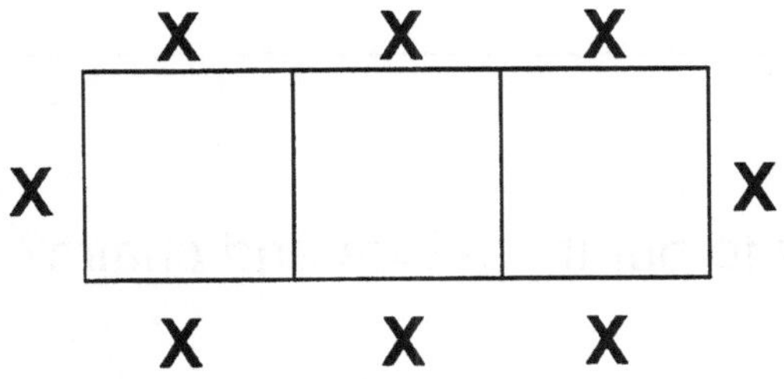

Put the **X** in these pictures:

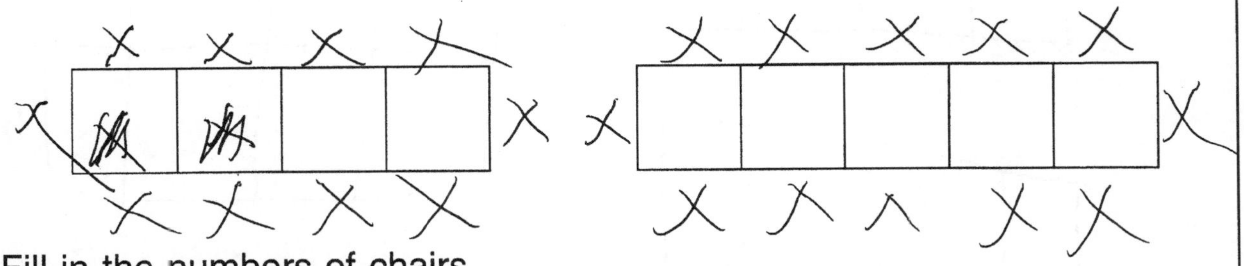

Fill in the numbers of chairs.

tables	chairs
1	4
2	6
3	8
4	10
5	12

We have used three different ways to put the chairs at the tables.

Which way do you like best? Why do you like it?

___

___

___

Which way means the most people can sit down?

___

___

___

Can you think of a different way to put the tables and chairs?

Draw it.

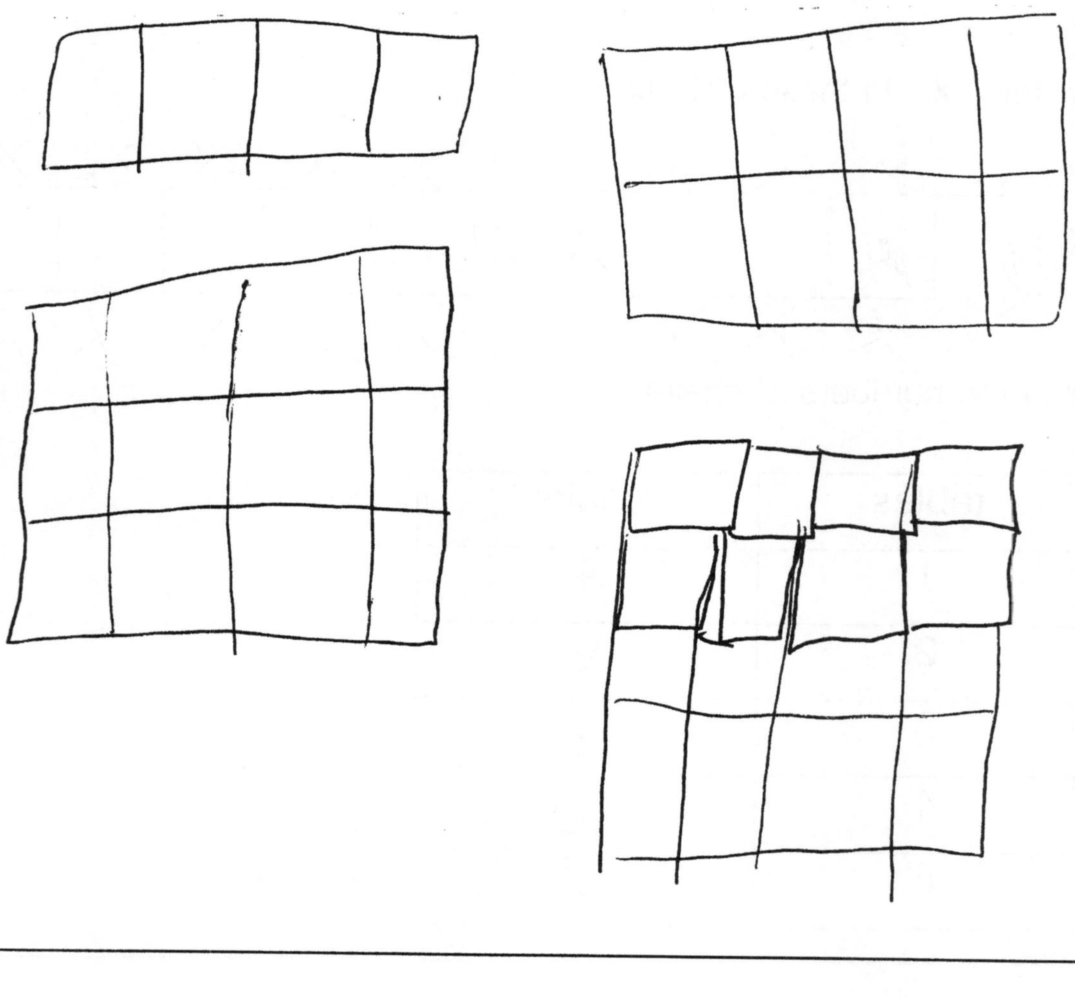

Year 1

Name: caroline

Tables and Chairs

The task

It is dinner time. We need to put the chairs at the tables.

We have to do it properly.

Like this:

X □ X   1 table has 2 chairs

X □ X

2 tables have 4 chairs

X □ X

X □ X

X □ X   3 tables have 6 chairs

X □ X

Draw the chairs.

Use X for a chair.

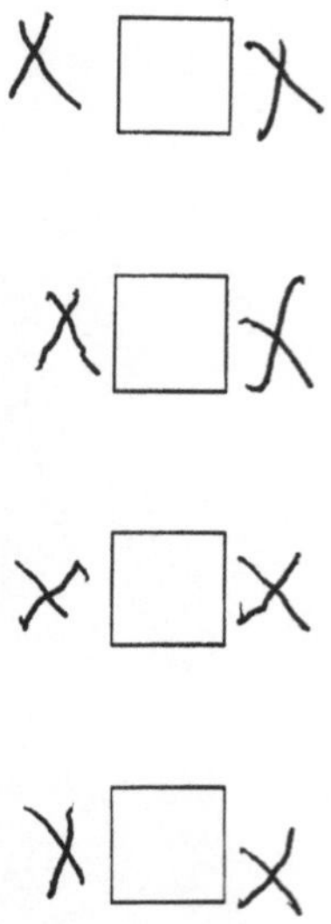

4 tables have 8 chairs

Draw 5 tables and the chairs.

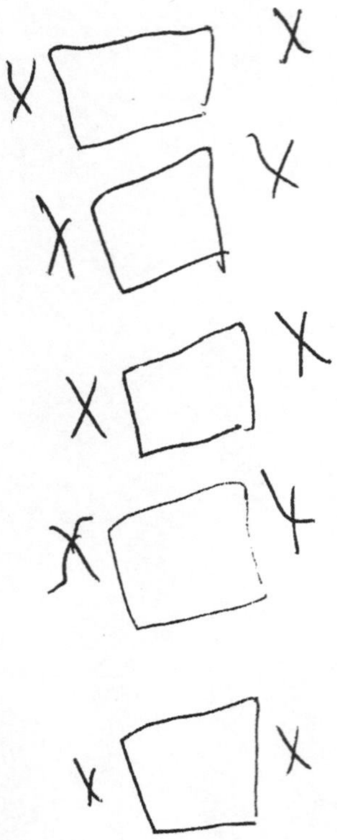

5 tables have 10 chairs

We can show our answers like this:

tables	chairs
1	2
2	4
3	6
4	~~16~~ 8
5	10
6	12
7	14
8	16

Can you write in the missing numbers of chairs?

You may want to draw pictures.

Or you may know the answer.

Can you see a pattern in the numbers?

Write it down. it is twos and doubles

Do you know how many chairs you need?

(a) 3 tables need 6 chairs

(b) 7 tables need 14 chairs

(c) 10 tables need 20 chairs

(d) 50 tables need 100 chairs

If you have 6 chairs how many tables are there? 3

x x x x x x

If you have 12 chairs how many tables are there? 6

More tables and chairs

More people come in for dinner.
We need to do the tables and chairs in a different way.

Like this:

1 table 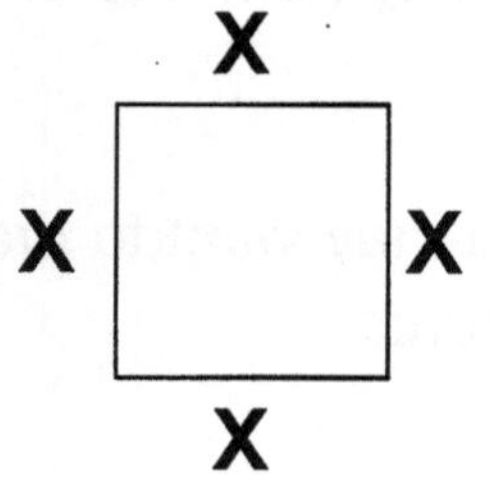

2 tables 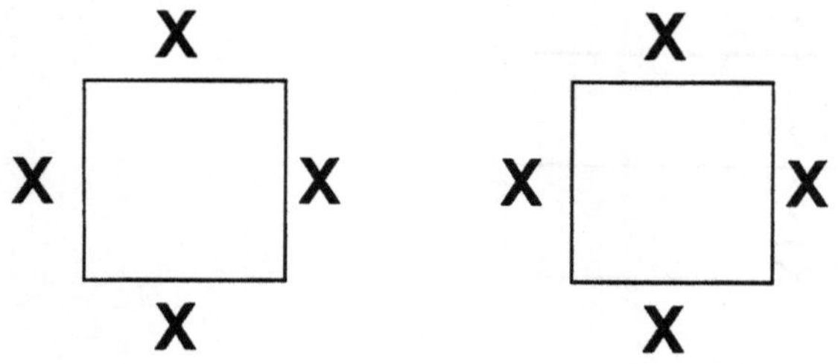

3 tables 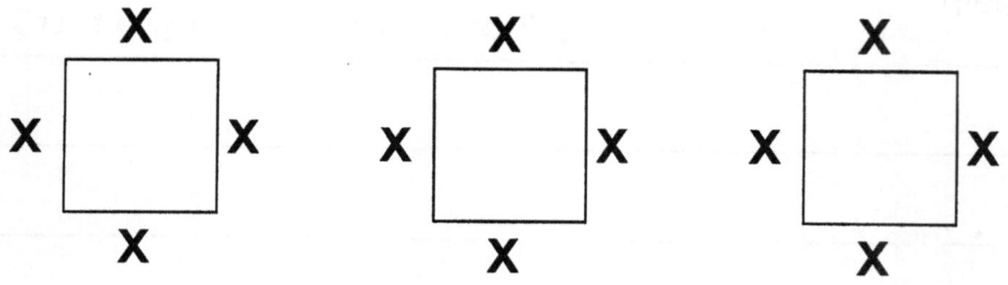

Fill in the numbers of chairs.

tables	chairs
1	4
2	8
3	12
4	16
5	20

You can draw pictures or you can work out the answer.

Can you see a pattern in the numbers? it is in 4s

Even more tables and chairs

We can push the tables together.

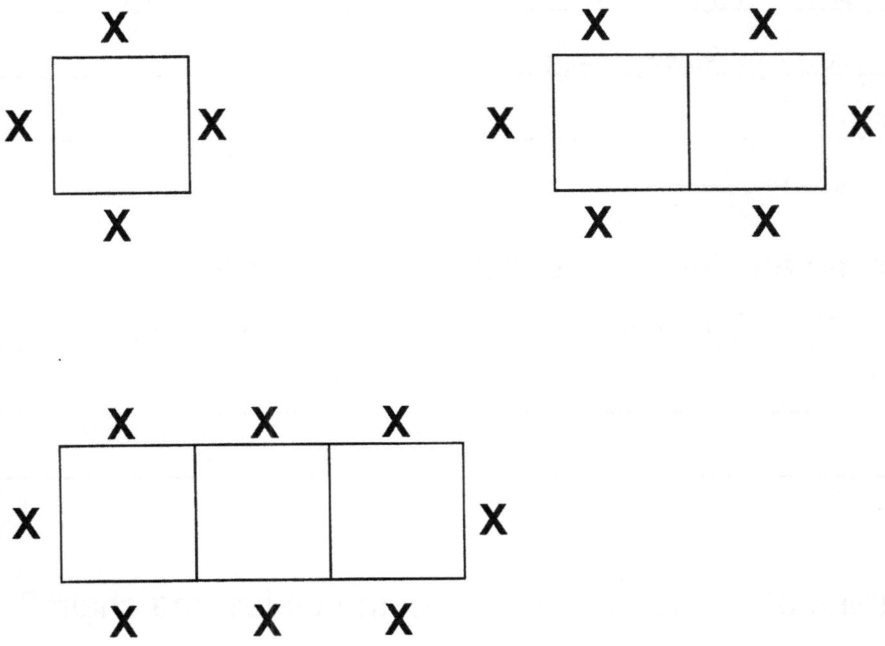

Put the X in these pictures:

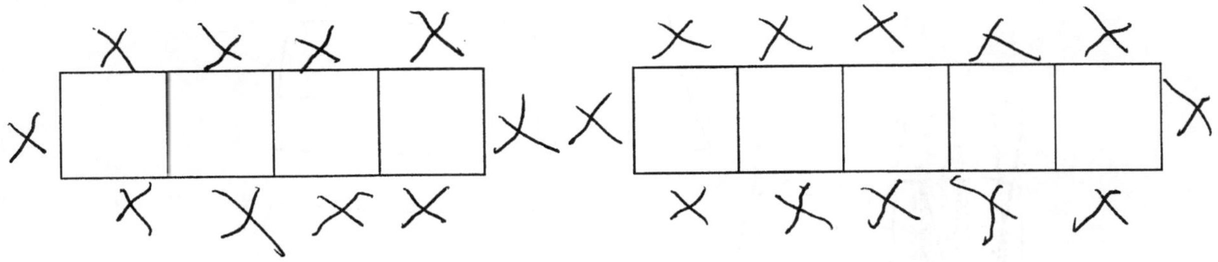

Fill in the numbers of chairs

tables	chairs
1	4
2	6
3	8
4	10
5	12

We have used three different ways to put the chairs at the tables.

Which way do you like best? Why do you like it?

I liked the one whar went
two four six

Which way means the most people can sit down?

the bigger is the fours

Can you think of a different way to put the tables and chairs?

Draw it.

4 POLYGONS AND POLYHEDRA

The task

If each corner is joined to every other corner investigate how many 'diagonals' there are in any polygon.

Introducing the task to your class

Prior to this investigation you may wish to spend some time looking at regular polygons or even constructing regular polygons. There are some excellent sets of polygons and polyhedra available from educational suppliers which the children can hold, study and compare.

It is quite useful to use elastic bands and pin boards for this investigation. The children should be very systematic when joining the corners, in order to see the pattern emerging. Introduce as much mathematical language as possible for example, vertices and the names of the polygons.

Square

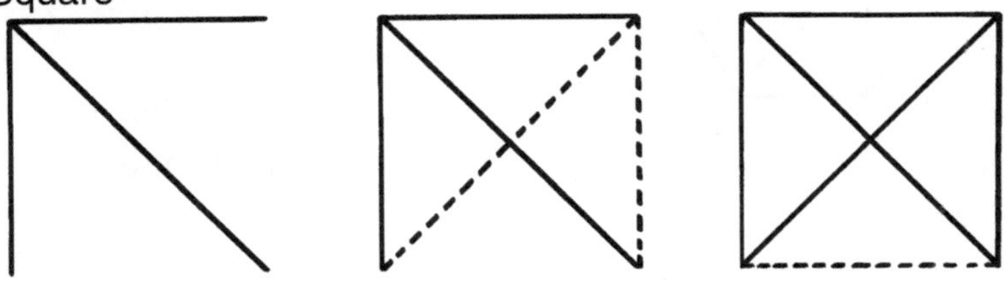

3 diagonals | 3 + 2 diagonals | 3 + 2 + 1 diagonals

Pentagon

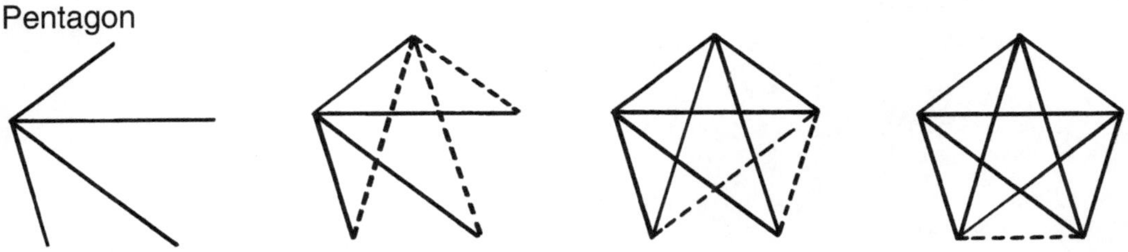

4 diagonals | 4 + 3 diagonals | 4 + 3 + 2 diagonals | 4 + 3 + 2 + 1 diagonals

No of sides, *s*	**Diagonals**	**No of diagonals, *d***
4	3 + 2 + 1	6
5	4 + 3 + 2 + 1	10
6	5 + 4 + 3 + 2 + 1	15

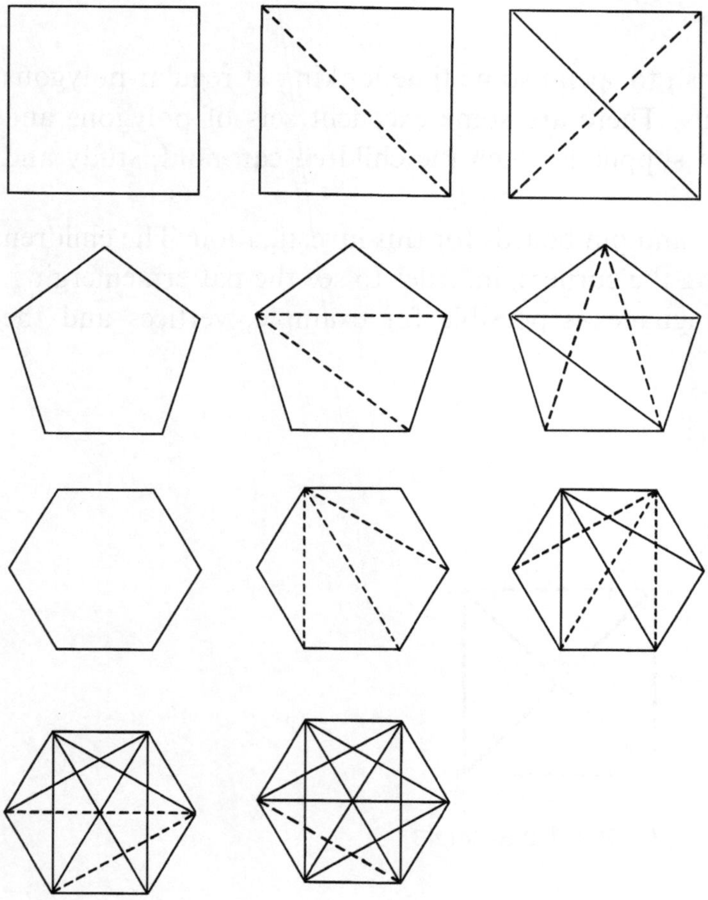

1 + 1
diagonals

2 + 2 + 1
diagonals

3 + 3 + 2 + 1
diagonals

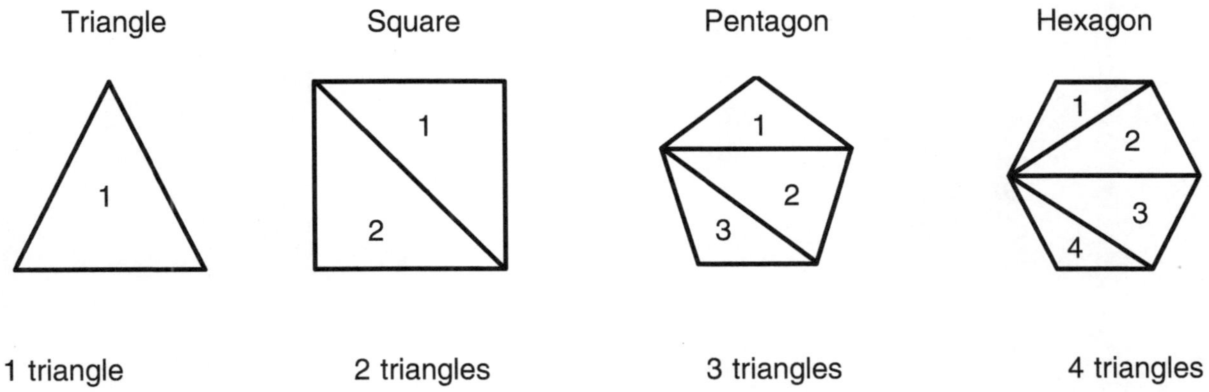

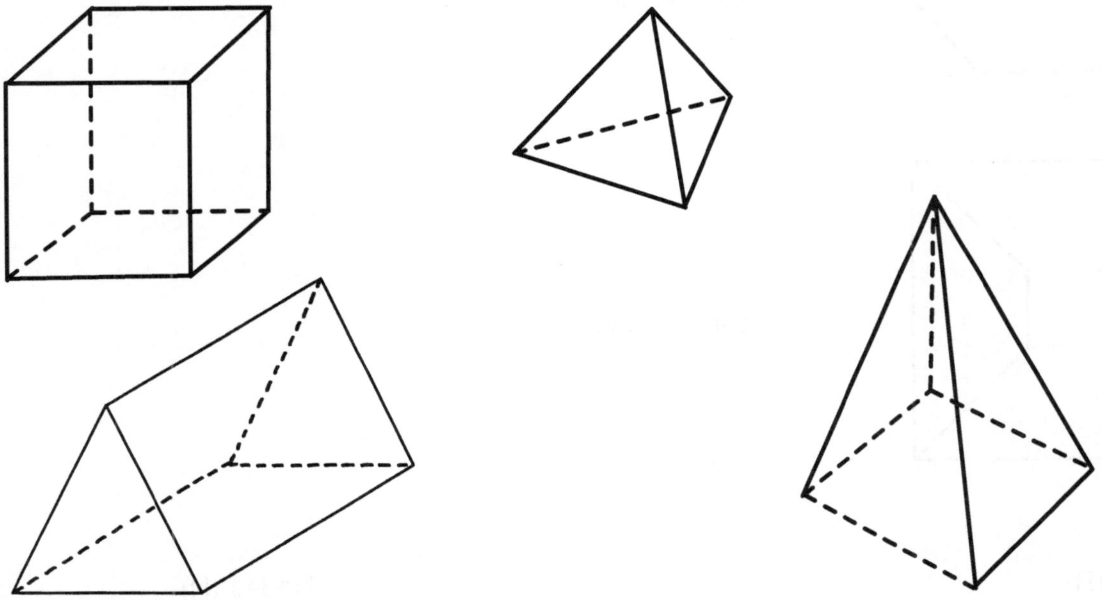

Shape	Number of faces, f	Number of vertices, v	Number of edges, e
Cube	6	8	12
Tetrahedron	4	4	6
Triangular prism	5	6	9
Square based pyramid	5	5	8

Cube

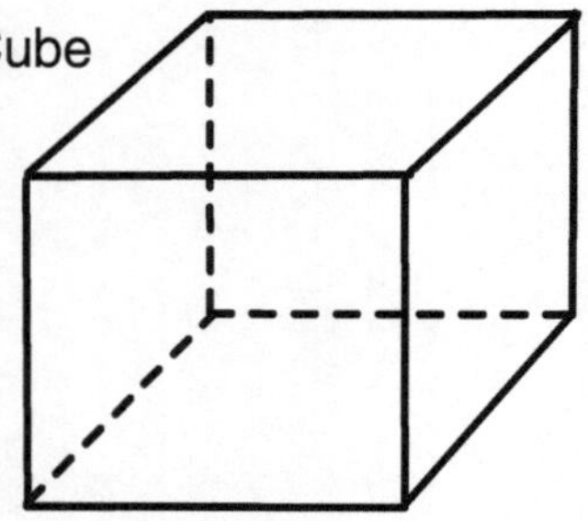

Skeleton model

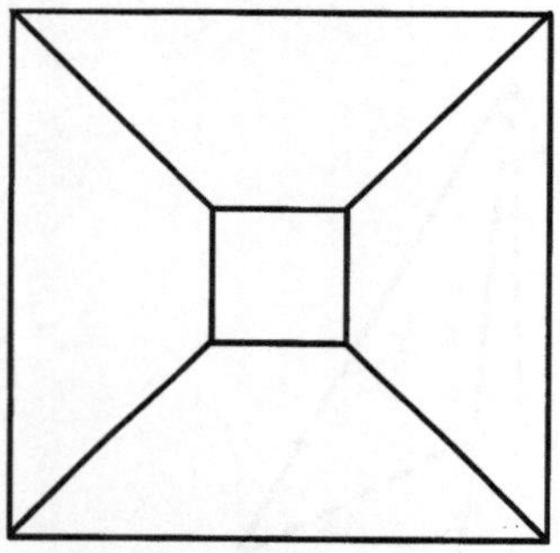

Schlegel diagram

Regions (R)

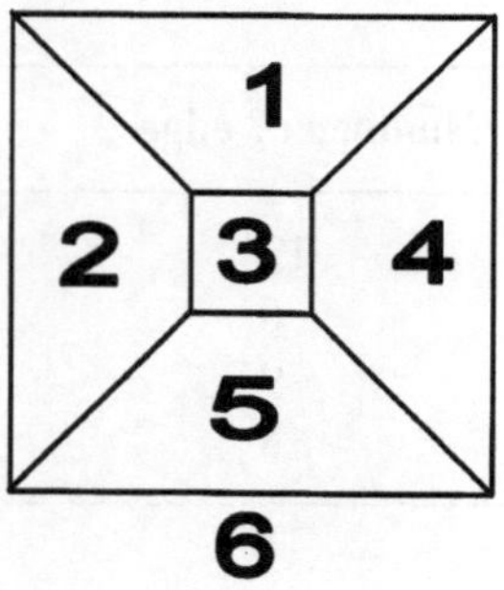

Arcs (A)

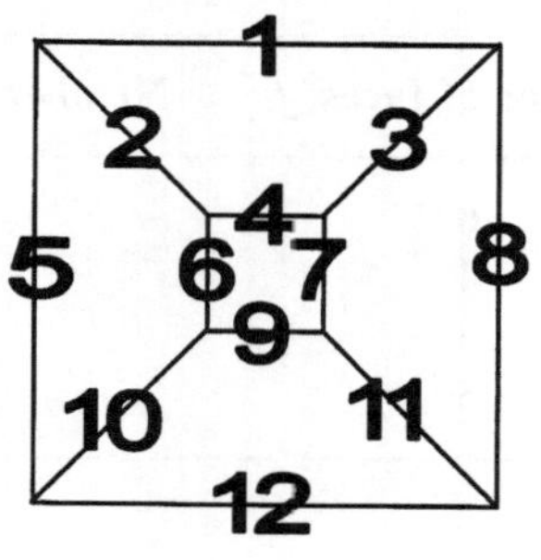

Nodes (N)

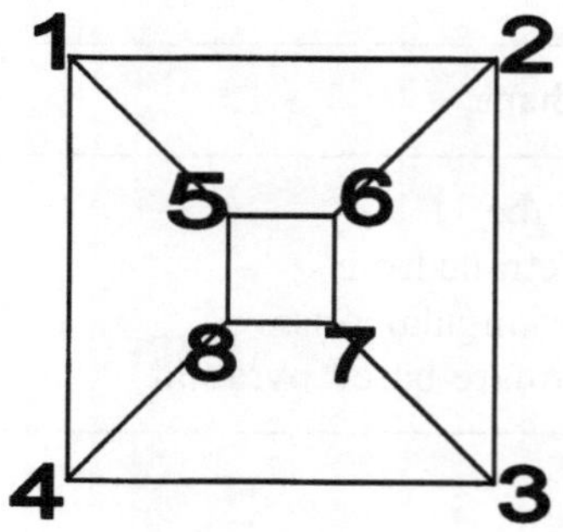

Shape	R	A	N
Cube	6	12	8

Square-based pyramid

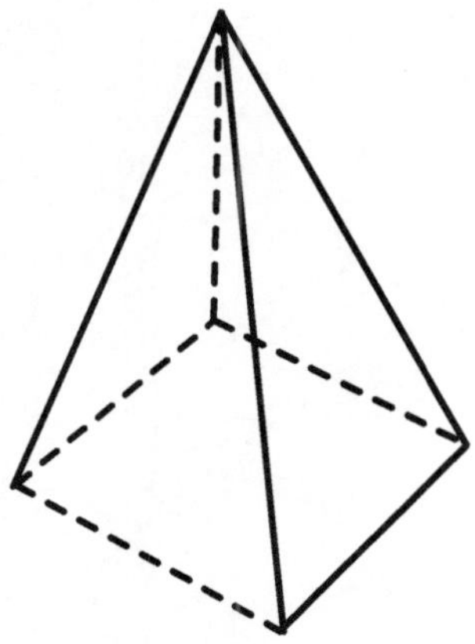

Skeleton model

Schlegel diagram with the square base removed

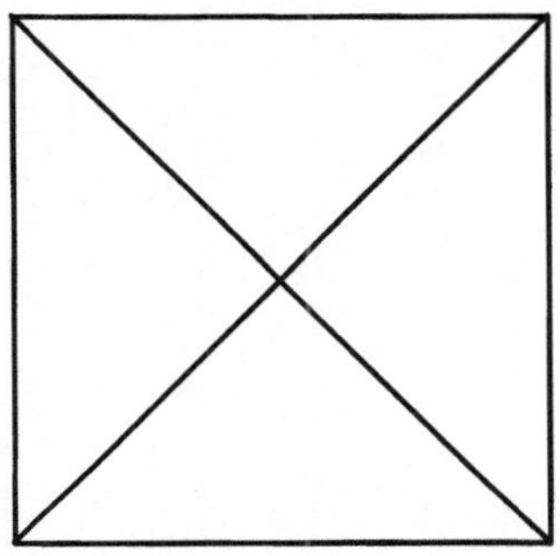

I

Schlegel diagram with a triangular face removed

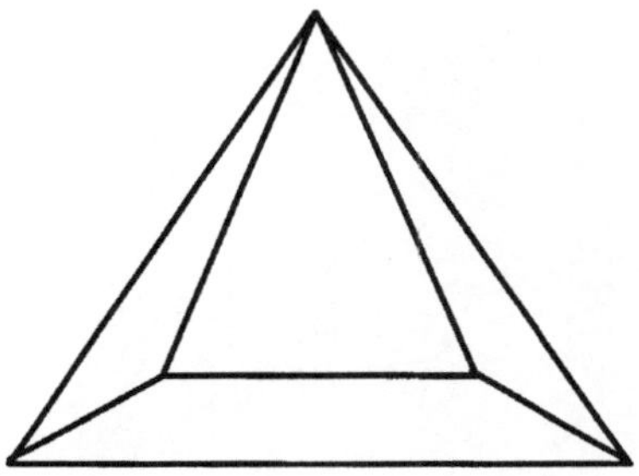

II

Square based pyramid	R	A	N
I	5	8	5
II	5	8	5

$$R + N = A + 2$$

5 ADDING TO 15

The task

List the numbers from 1 to 9.

Player 1 plays against player 2.

Each takes it in turn to choose a number from the list. This number is then crossed out.

The aim is to be the first to add their numbers up to 15. Not every number they have chosen needs to be used in the sum; as seen below player 1 does not use the number 3 to make the total of 15.

Introducing the task to your class

Introduce this investigation by playing the game: you against the class.

Later we will consider if there is a strategy you can adopt to increase your chances of winning.

Player 1 = Alex

Player 2 = James

Below shows what happens, step by step, in a game.

Alex	James	Current state of numbers list	Comment
		1 2 3 4 5 6 7 8 9	
3			Three is now missing from the list of possible numbers to choose from.
		1 2 4 5 6 7 8 9	With 9 James now needs 6 to reach 15 so Alex must choose 6 to block him.
	9		
		1 2 4 5 6 7 8	
6			
		1 2 4 5 7 8	
	1		
		2 4 5 7 8	
5			
		2 4 5 7 8	Alex could make 15 using 6 + 5 then 4, so James chooses 4 to block.
	4		
		2 7 8	
7			Alex wins: 7 + 5 + 3 = 15 NOTE: not all chosen numbers need to be used to make 15.

Writing up the investigation

The maths behind the game.

Some numbers are more useful to choose than others.

Pupils list the sums that add to 15. This is quite tricky. Pupils need to be taught how to list systematically.

Initially limit the sums to those that contain two numbers then three numbers and build up.

6,9 7,8

1,5,9 1,6,8 *1,7,7*

> !! this cannot be included as it uses the number 7 twice, and this is useful for pupils to note

Now move onto sums that start with 2

2,4,9

> !! discuss with pupils how they know what the first sum is, i.e. why not 2,3

2,5,8 2,6,7 3,4,8 3,5,7 4,5,6

Now move onto addition sums with four numbers in them

1,2,3,9 1,2,4,8 1,2,5,7

1,3,4,7 1,3,5,6 2,3,4,6

number	**frequency**
1	7
2	7
3	6
4	6
5	6
6	6
7	5
8	5
9	4

In a discussion, talk about how the information they have collected may inform their choice of numbers.

Extension ideas

To extend this task you may increase the choice of numbers in your starting list, perhaps from 1 to 12.

Again encourage the pupils to discuss what they think may happen before it does.

Ask pupils to adopt a certain strategy and monitor how they do.

6 HIDDEN FACES

The task

When a die in the shape of a cube is rolled, one of the faces is always covered and the other five faces are visible.

What is the largest possible total of all the faces visible when one die is rolled?

What is the largest possible total of all the faces visible when two dice are rolled?

What is the largest possible total of all the faces visible when three dice are rolled?

What is the largest possible total of all the faces visible when n dice are rolled?

Introducing the task to your class

Children love using different shaped dice and various sizes of dice, so try to have as many different types available as possible, although for this investigation we will initially be using cubic dice. You may also wish to have nets of cubes available.

If you have a very large die you can hold the die against your chest and ask the children to find the total of the faces which they can see. Throw the die to a pupil and then ask the rest of the class to repeat the exercise; this can be great fun.

Writing up the investigation

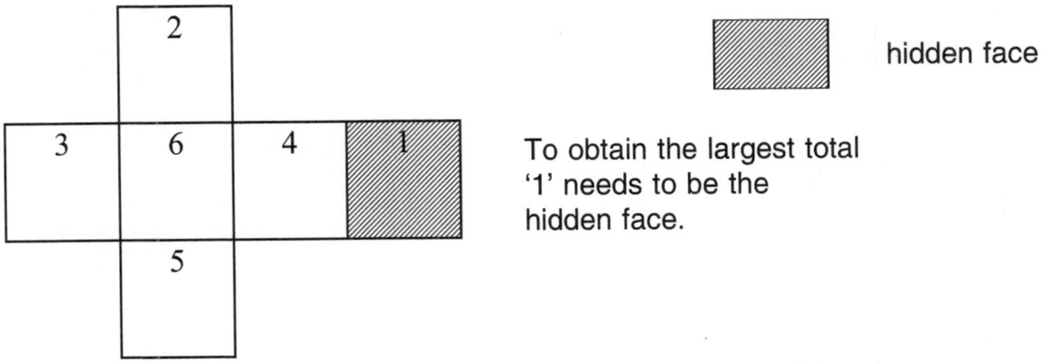

To obtain the largest total '1' needs to be the hidden face.

Pupils should quickly establish that opposite faces on the die add up to 7.

No of dice	Total	
1	$2+3+4+5+6$	20
2	$2(2+3+4+5+6)$	40
3	$3(2+3+4+5+6)$	60
n	$n(2+3+4+5+6)$	$20n$

$t = 20n$ where n is the number of dice and t is the largest possible total.

Extension ideas

What is the largest possible total if two dice are rolled and the two dice land next to each other?

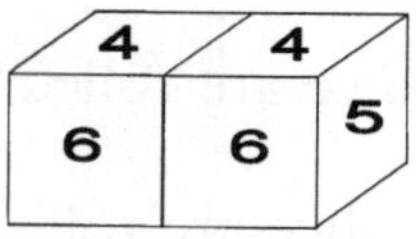

What is the largest possible total if three dice are rolled and the dice land next to each other?

What is the largest possible total if four dice are rolled and the dice land next to each other?

What is the largest possible total if n dice are rolled and the dice land next to each other?

Two dice

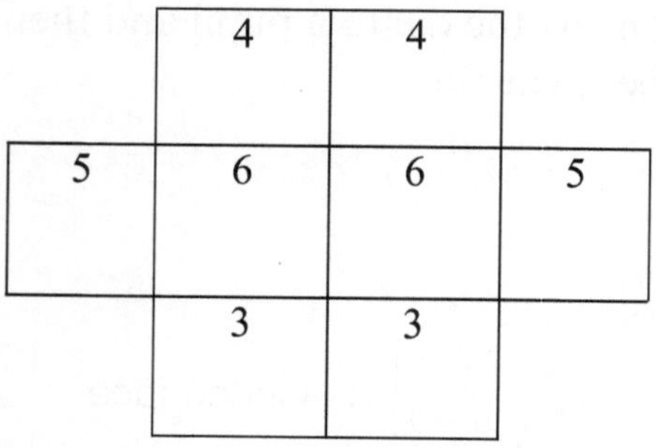

Three dice

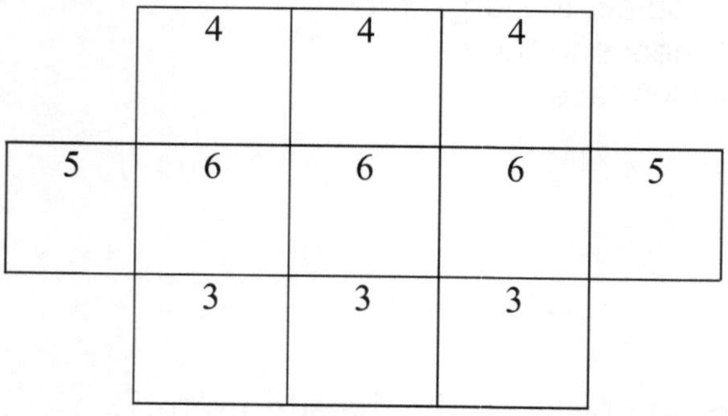

Four dice

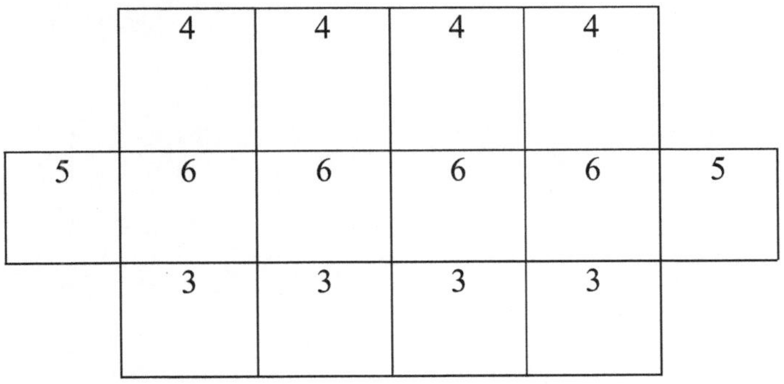

Number of dice side by side	Total
2	36
3	49
4	62
n	$13n + 10$

$$t = 13n + 10$$

Try to encourage your pupils to explain why the rule is $13 \times n + 10$.

The 'plus 10' is because the two ends are always 5 and 5.

The 13 is because each column is $4 + 6 + 3$.

What happens to the total if the dice make an L shape?
What happens to the total if the dice are in a tower?

Investigate the totals with other types of polyhedron dice.

7 DOTTY INVESTIGATION

Introducing the task to your class

Start the activity by drawing out a large square grid of dots in chalk on the floor or set out 25 chairs in a 5 by 5 grid. Using the pupils, get them to arrange themselves in the smallest Z arrangement. Gradually increase the arrangement by adding in more pupils. (Pupils watching will probably find it easier to see the spatial pattern if the pupils are seated.)

When recording, pupils may then find it easier to draw the shapes on square spotty paper rather than squared paper.

Pupils may want to set out counters to see the physical arrangement before they draw the shapes.

By being physically involved, pupils will soon see that to grow the shape each time involves adding in three extra pupils. This is not as apparent simply from drawing the shape.

Writing up the investigation

You will see from the included pupil's piece of work how the shapes build up.

As the pupil starts the second shape arrangement, the 'T', she notices that the dots increase by four. This is a visual observation, which she records in a *thought bubble* at the time of observation. This chronological recording is vital. As a teacher you need to see **when a piece of thinking has occurred**.

To encourage this way of working you could give stickers as a reward for when a pupil uses a thought bubble.

As you can see the, pupil then feels confident enough to bypass the 'down-the-table' addition rule and go straight to the 'across-the-table' rule.

Very little teacher input was given during this stage. The pupil worked independently, happy to use the earlier Z investigation as a model.

Dotty Investigation

The Task

Look at the shapes below. They are made from joining dots.

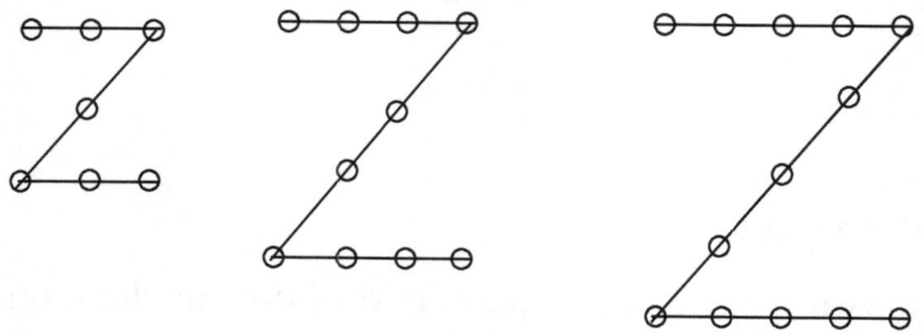

Can you see how the shapes grow?

Copy the three Z shapes and draw the next three in the pattern.

Count how many dots are in each pattern and record your results in a table.

Shape number *s*	Number of dots *d*
1	7
2	
3	
4	
5	

Can you find a pattern in the number of dots?

Can you predict how many dots will be in the 7th shape and the 10th shape?

Try to find a rule that links the shape number and the number of dots.

Can you use your rule to predict the number of dots in the 100th shape?

Can you write a formula?

Now investigate the following dotty shapes:

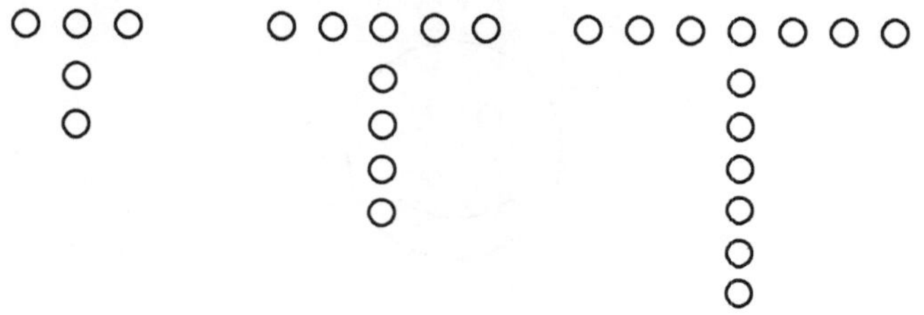

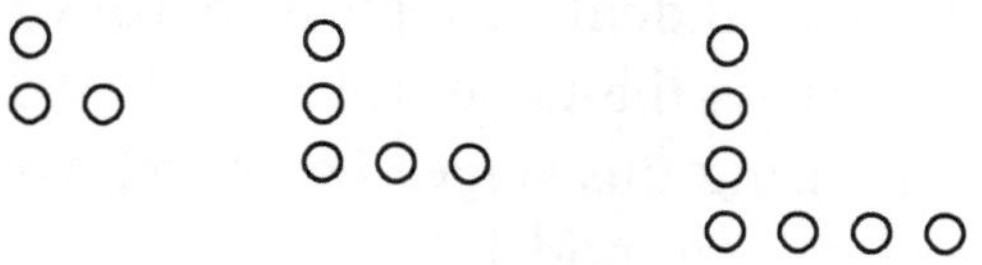

A Year 2 pupil

Dotty Investigation

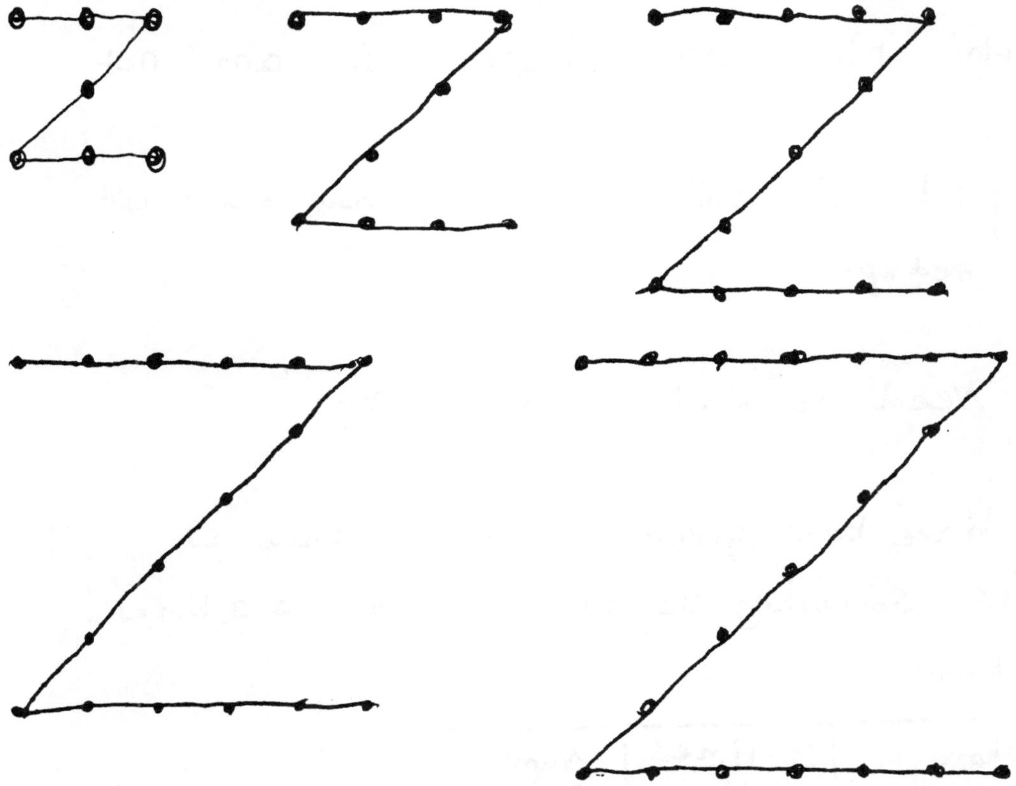

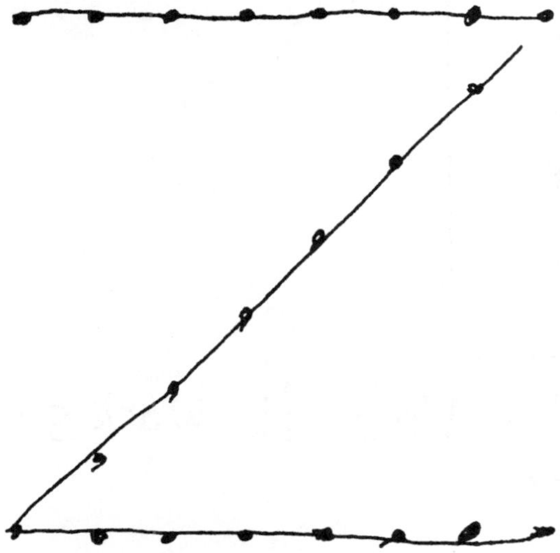

shape number	Number of dots	
1	7	+3
2	10	+3
3	13	+3
4	16	+3
5	19	+3
6	22	+3
7	25	+3
8	28	+3
9	30	+3
10	34	+3

My teacher has asked me to Find the 100th shape. I am not Happy !!! It will take a long time and I will get fed up.

I Need to Find a quicker way.

I Have been given a clue the clue is its Something to do withe the 3 times table.

shape Number		3 times table		Number of dots
1	×3 =	3	+4 →	7
2	×3 =	6	+4 →	10
3	×3 =	9	+4 →	13
4	×3 =	12	+4 →	16
5	×3 =	15	+4 →	19

10 ×3 = 30 +4 → 34 Yes it Works!

100 ×3 = 300 +4 → 304

If I know the shape Number I Multiply by 3 then add 4 ~~wich~~ ~~wit~~ Which is the Number of dotts.

$S \times 3 + 4 = D$

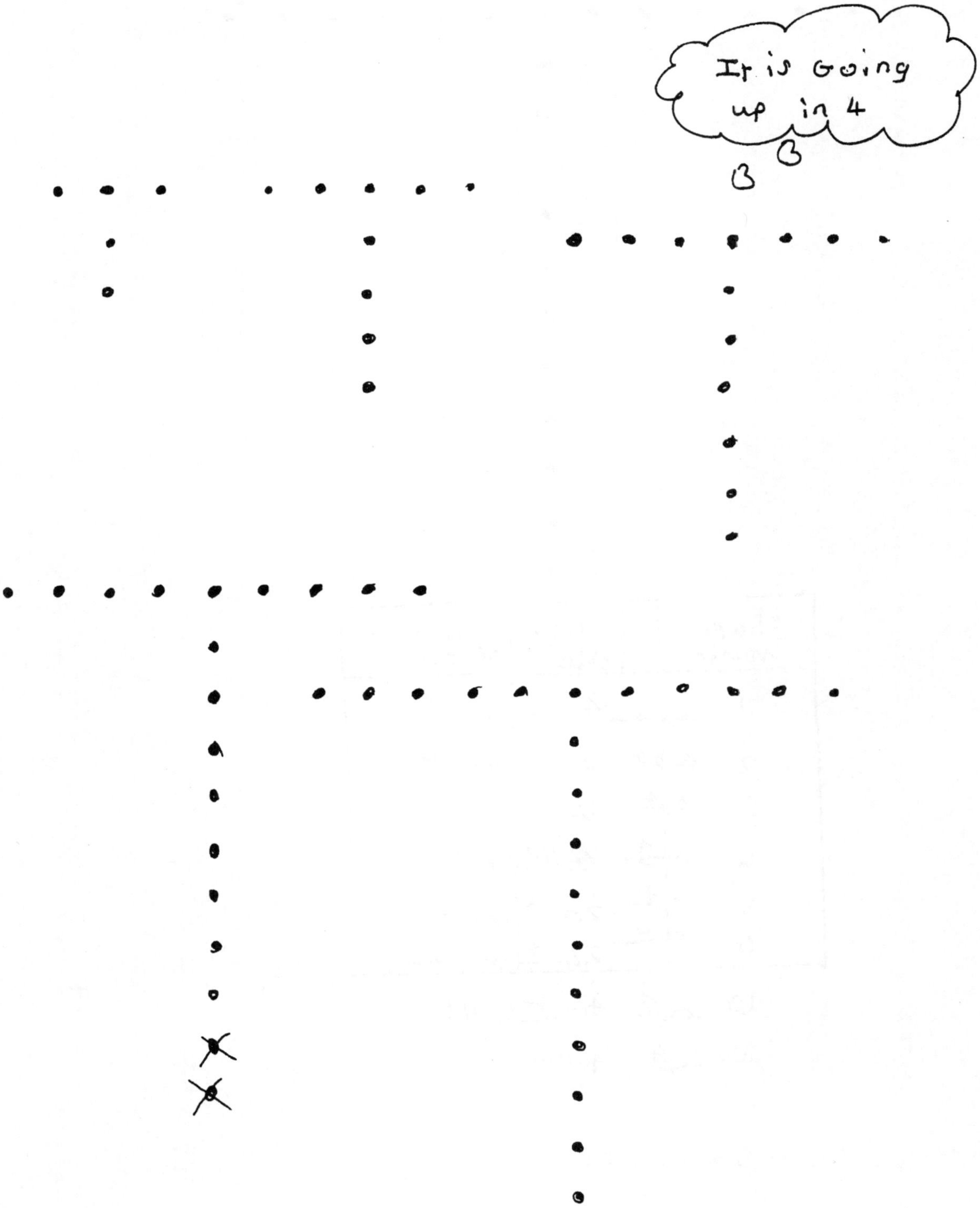
It is Going
up in 4

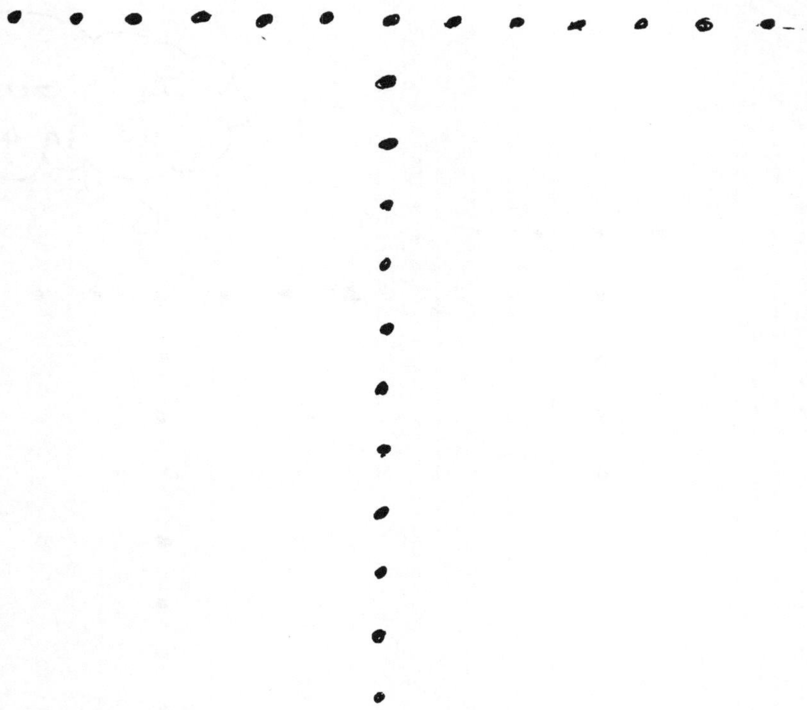

Shape number	4 times table	Number of dotts
1 ×4 →	4 +1 →	5
2 ×4 →	8 +1 →	9
3 ×4 →	12 +1 →	13
4 ×4 →	16 +1 →	17
5 ×4 →	20 +1 →	21
6 ×4 →	24 +1 →	24

10 ×4 → 40 +1 → 41

100 ×4 → 400 +1 = 401

S ×4 + 1 = □.

Extension ideas

Later, when trying to relate the formula to the arrangement, you need to distinguish the corner pupils from the others. This should help them to see how the shape grows and how this relates to their formula, as shown below.

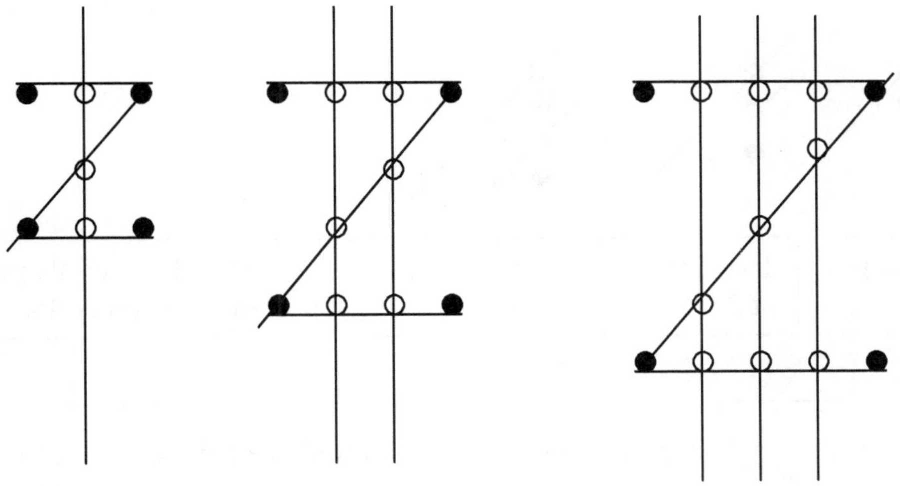

One group of 3

Two groups of 3

Three groups of 3

Then add on the 4 corner dots.

The formula is

shape number $\times$ 3 + 4 = number of dots

or

$$s \times 3 + 4 = d$$

or

$$3s + 4 = d$$

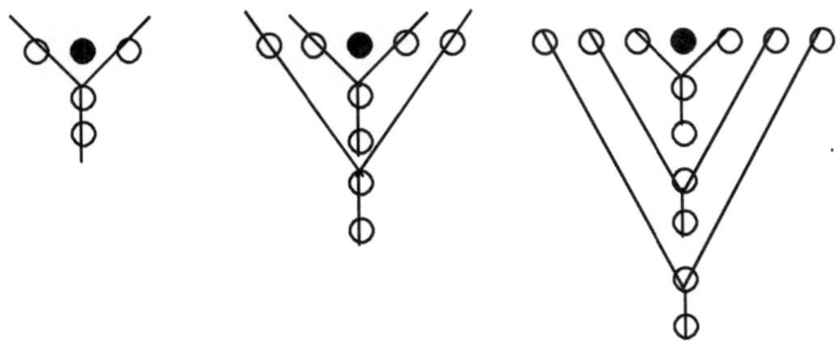

One group of 4

Two groups of 4

Three groups of 4

For each then add on the top centre dot

Formula

$$\text{shape number} \times 4 + 1 = \text{number of dots}$$
$$s \times 4 + 1 = d$$
$$4s + 1 = d$$

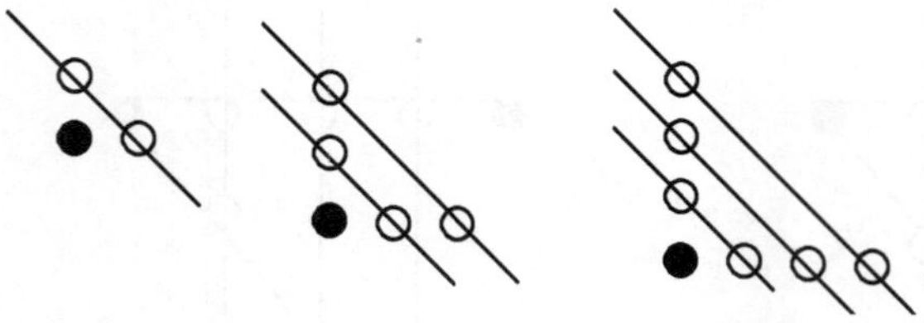

One group of 2	Two groups of 2	Three groups of 2	Then for each shape add on the one corner dot

Formula

$$\text{shape number} \times 2 + 1 = \text{number of dots}$$
$$s \times 2 + 1 = d$$
$$2s + 1 = d$$

!! If pupils are confident they only need to write down $2s + 1 = d$; the other lines are not necessary. However to achieve the highest level they do need to relate the formula to the dots arrangement as shown.

8 POND BORDERS

The task

Katie has started working in a garden centre that sells ponds. The ponds are made to a special design and a path surrounds each of the ponds. The paving slabs used in the path measure 1 ft by 1 ft. Katie works every Saturday and Saturday mornings are very busy.

Katie would really like to know a quick way of working out how many slabs are needed to surround any size of pond.

How many paving slabs are needed for the path that surrounds a pond measuring

(i) 3 ft by 3 ft? (ii) 20 ft by 20 ft? (iii) p ft by p ft?

Introducing the task to your class

Start the task by investigating square ponds. We have used carpet squares for this and the children really get involved in the task. You will need plenty of room if you want the children to build up the sizes of the ponds and leave the ponds on display. If you are unable to get hold of any sort of tiling then use multilink or cm cubes. You will find differentiated worksheets that you may wishto use at the end of this activity.

Writing up the investigation

Encourage the children to be systematic. Draw a 1×1 pond and then surround the pond with square paving slabs. Continue with 2×2, 3×3 etc. They should try to predict the number of paving slabs needed for a 4×4 pond or the 5×5 pond.

When the pupils have found the formula for any size of square pond they should try to explain where the formula comes from.

p is the size of the pond

s is the number of slabs

$$4p + 4 = s$$

Four times the size of the pond because of the four sides

Plus four for the four corners

Extension ideas

This investigation can easily be extended to investigate rectangular ponds, where the rule is very similar to the one for square ponds.

If the rectangular pond is a ft by b ft then the rule will be

$$2a + 2b + 4 = s$$

Two sides of length a — Two sides of length b — Four corners

You might like to consider investigating ponds in the shape of their initials. There are families of shapes that have similar rules. It is interesting to consider why one particular shape has exactly the same rule as another.

Consider surrounding the square ponds with layers of paving slabs for example.

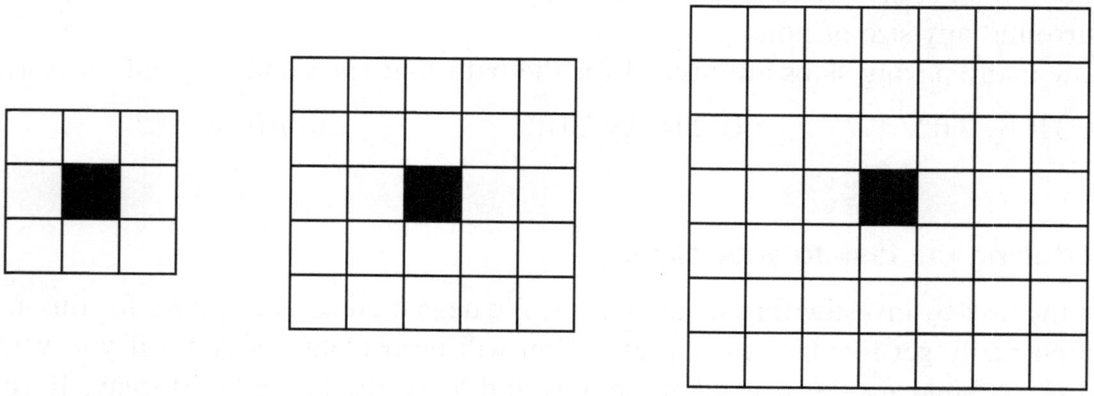

1 layer   2 layers   3 layers

No of layers squared

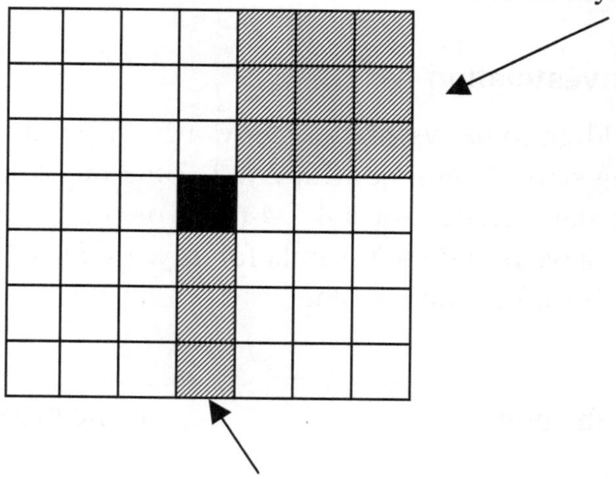

No of layers × pond size

If n is the number of layers

$$4n^2 + 4np = s$$

For a rectangular pond with n layers

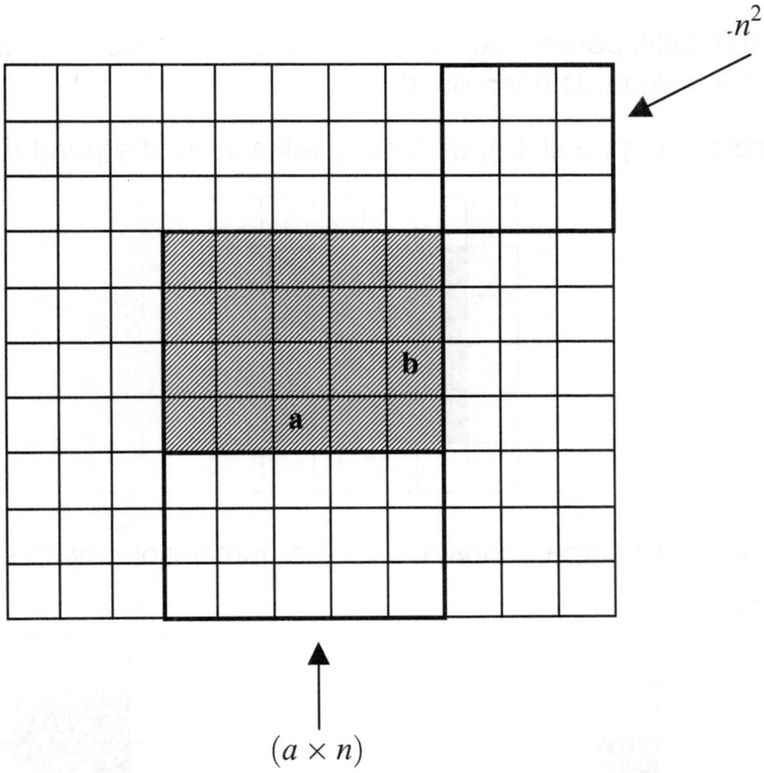

$$4n^2 + 2(an) + 2(bn) = s$$

Pond Borders (1)

A garden centre sells ponds that it makes to a special design. A path made of paving slabs 1 ft by 1 ft surrounds each pond.

E.g. a path containing 20 slabe, as below, will surround a pond that is 4 ft by 4 ft:

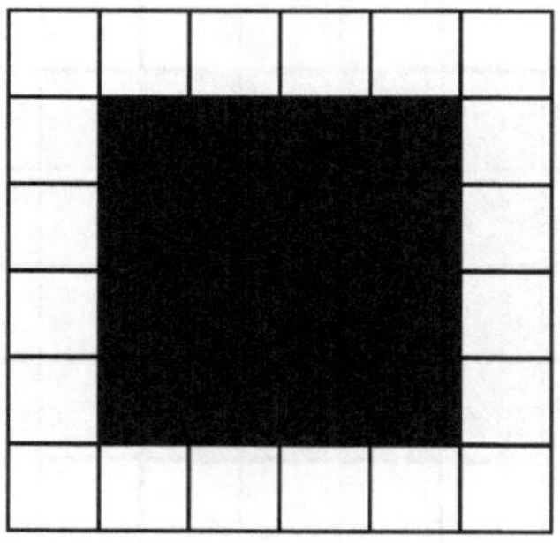

Below are pictures of some ponds. Count the number of paving slabs needed.

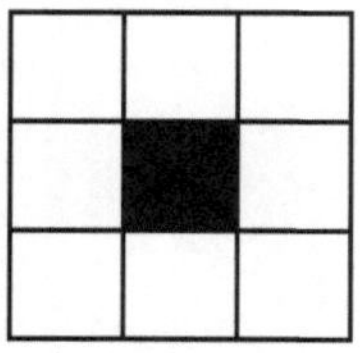

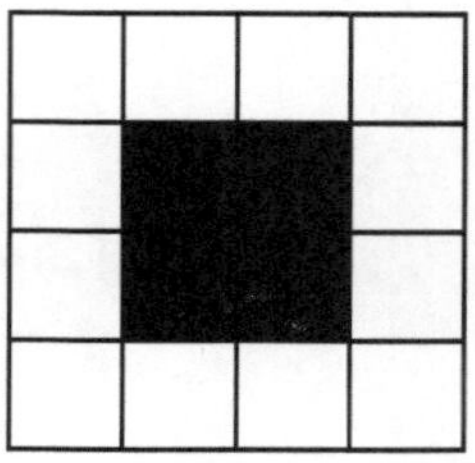

Draw ponds that are 3 ft by 3 ft then 5 ft by 5 ft and 6 ft by 6 ft.

Put your results into a table as below.

What happens to the number of slabs each time the size of the pond increases?

Size of pond (in feet)	Number of slabs
1 × 1	8
2 × 2	
3 × 3	
4 × 4	20
5 × 5	
6 × 6	

Use this rule to predict the number of slabs needed for ponds up to 10 ft by 10 ft and fill in the table.

Can you find a quicker way of working out the number needed for a pond 20 ft by 20 ft?

Test your idea by drawing the pond.

Try to find a rule that can be used for any size pond.

What about rectangular ponds? Try to use a good system to investigate this new shape.

Pond Borders (2)

A garden centre sells ponds that it makes to a special design. A path made of paving slabs 1 ft by 1 ft surrounds each pond.

E.g. a path containing 20 slabs, as below, will surround a pond that is 4 ft by 4 ft

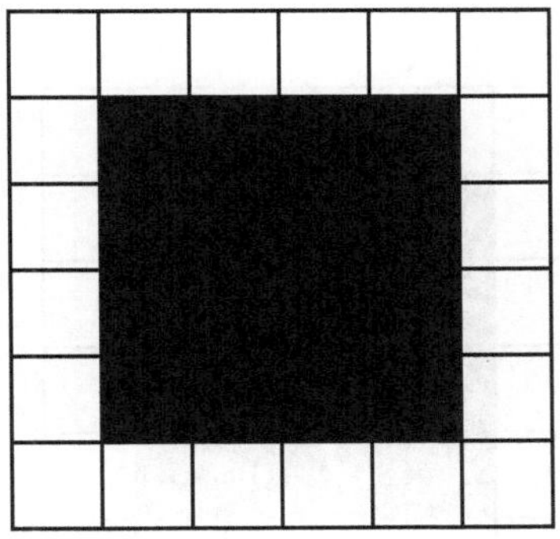

Draw ponds that are 1 ft by 1 ft, 2 ft by 2 ft, etc. Count the number of paving slabs needed.

Size of pond (in feet)	Number of slabs
1 × 1	8
2 × 2	

Put your results into a table as below.
What happens to the number of slabs each time the size of the pond increases?

Use this rule to predict the number of slabs needed for ponds up to 10ft by 10ft and fill in the table.

Can you find a quicker way of working out the number needed for a pond 20 ft by 20 ft?

Test your idea by drawing the pond and try to find a rule that can by used for any size pond.

Investigate for other shape ponds (keep your shapes simple and try to plan a system).

Pond Borders (3)

A garden centre sells ponds that it makes to a special design. A path made of paving slabs 1 ft by 1 ft surrounds each pond.

E.g. a path containing 20 slabs, as below, will surround a pond that is 4 ft by 4 ft.

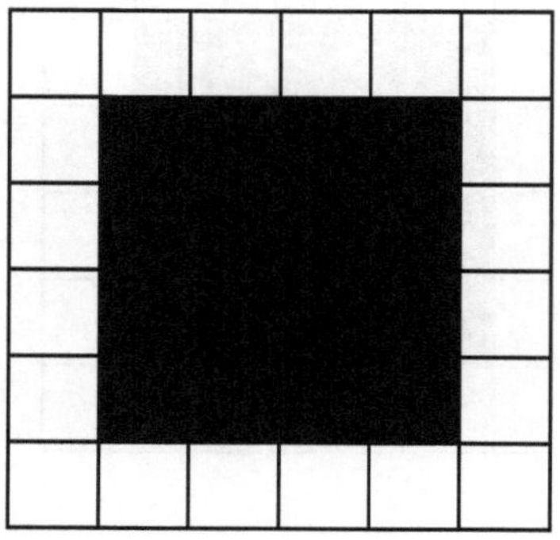

Investigate for other size square ponts and tabulate your results.

Remember to make predictions and test them.

Look for patterns and rules to connect the size of the pond to the number of slabs needed.

Try to write your rules in words then use ‘s’ for size of pond and ‘n’ for the number of slabs. Try to use algebra to write your rule.

Oliver Year 5

Pond Borders

I work at a garden centre and we sell square ponds but all the people that come to buy them whant parving slabs to go aroud.

Ponds are measured in feet, 1 paving slab is 1ft x 1ft

my aim is to find quickly how many slabs ~~for~~ any size pond I need to start with the simplest pond and bild up systematically I will start 1x1 2x2 3x3 4x4

1x1 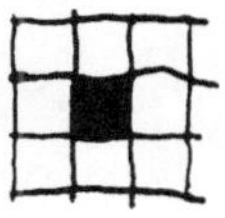 8 slabs

2x2 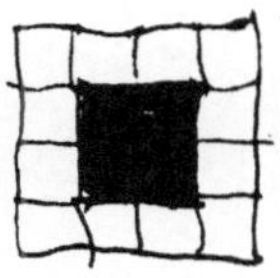 12 Slabs

3x3 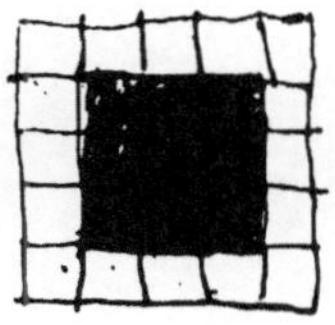 16 Sabs

Will it stay +4 so it might be 16 then 20 then 24

Size of pond	Slabs
1x1	8
2x2	12
3x3	16
I prodict 4x4	20

4X4  20 Yipee! I'm right 

as long as people want small ponds itsok

But is you want 50X50 it will take ages

we neeld to sind a quicker methoned

Size of pond, p	X4	+4 Slabs, s
1 X 1	4	8
2 X 2	8	12
3 X 3	12	16
4 X 4	16	20
5 X 5	20	24

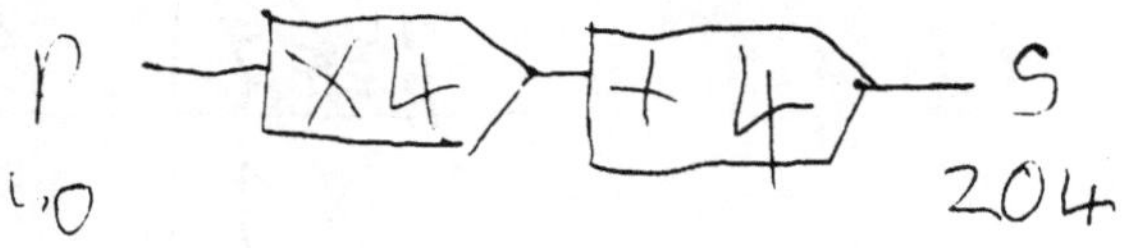

$p4 + 4 = s$

$4p + 4 = s$

Explaining the formula

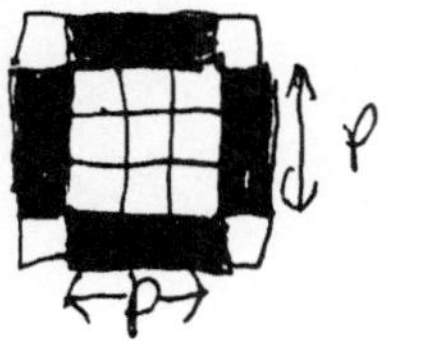

there are 4 sides so we need 4 more for the corners

Rectangular ponds

we are going to use our knowlege about the square ponds to help us.

Both types of pond have 4 sides and corners

The difference is rectangles have 2 pairs of sides

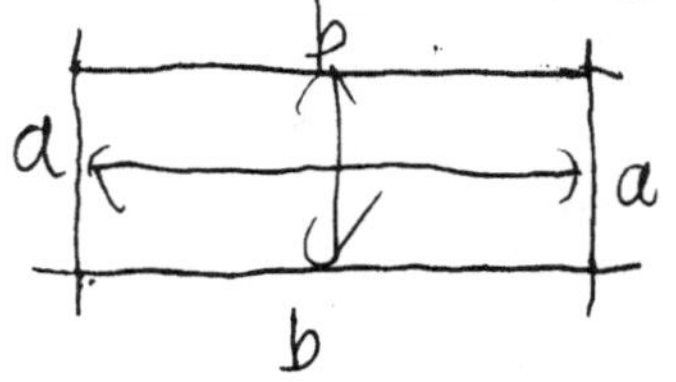

I think the formula is

S = 2×a + 2× B + 4

S=2a + 2B+4= Yes

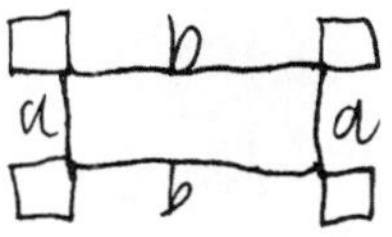

S= 2×2+2×17+4
= 4+34+4
= 42

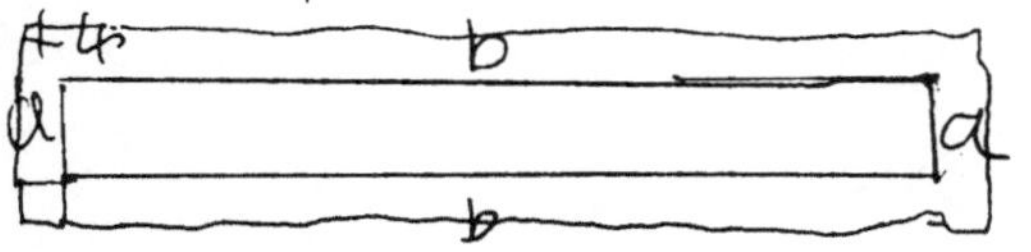

42

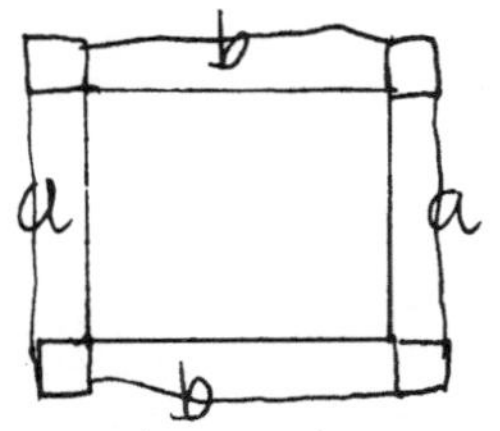

S= 2×6×2×5
= 12+10+4
= 26

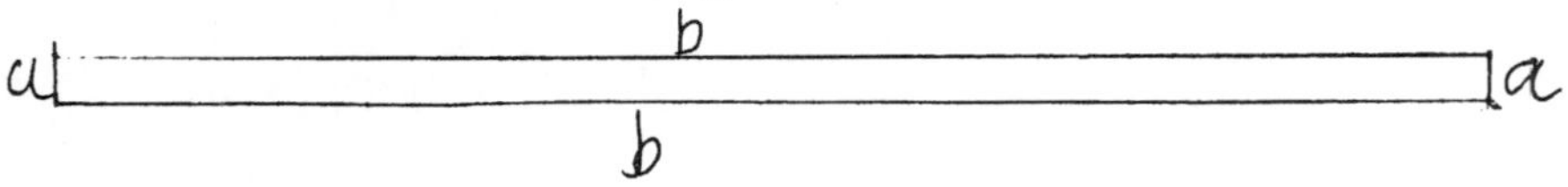

S=2×1+2×32+4
= 2+64+4
= 70

Yes